THE HISTORY

OF THE

ROYAL AGRICULTURAL SOCIETY

OF ENGLAND

1839—1939

by

PROFESSOR J. A. SCOTT WATSON

British Library Cataloguing-in-Publication Data
A catalogue record for this book is available from
the British Library

Farming

Agriculture, also called farming or husbandry, is the cultivation of animals, plants, or fungi for fibre, bio-fuel, drugs and other products used to sustain and enhance human life. Agriculture was the key development in the rise of sedentary human civilization, whereby farming of domesticated species created food surpluses that nurtured the development of civilization. It is hence, of extraordinary importance for the development of society, as we know it today. The word *agriculture* is a late Middle English adaptation of Latin *agricultūra*, from *ager*, 'field', and *cultūra*, 'cultivation' or 'growing'. The history of agriculture dates back thousands of years, and its development has been driven and defined by vastly different climates, cultures, and technologies. However all farming generally relies on techniques to expand and maintain the lands that are suitable for raising domesticated species. For plants, this usually requires some form of irrigation, although there are methods of dryland farming. Livestock are raised in a combination of grassland-based and landless systems, in an industry that covers almost one-third of the world's ice- and water-free area.

Agricultural practices such as irrigation, crop rotation, fertilizers, pesticides and the domestication of livestock were developed long ago, but have made great progress in the past century. The history of agriculture has played a major role in human history, as agricultural progress has been a crucial factor in worldwide socio-economic change. Division of labour in agricultural societies made (now) commonplace specializations, rarely seen in hunter-gatherer cultures, which allowed the growth of towns and cities, and the complex societies we call civilizations. When farmers became capable of producing food beyond the needs of their own families, others in their society were freed to devote themselves to projects other than food acquisition. Historians and anthropologists have long argued that the development of agriculture made civilization possible.

In the developed world, industrial agriculture based on large-scale monoculture has become the dominant system of modern farming, although there is growing support for sustainable agriculture, including permaculture and organic agriculture. Until the Industrial Revolution, the vast majority of the human population laboured in agriculture. Pre-industrial agriculture was typically for self-sustenance, in which farmers raised most of their crops for their own consumption, instead of cash crops for trade. A remarkable shift in agricultural practices has occurred over the past two centuries however, in response to new technologies, and the development of world markets. This also has led to technological improvements in agricultural techniques, such as the Haber-Bosch method for synthesizing ammonium nitrate which made the traditional practice of recycling nutrients with crop rotation and animal manure less important.

Modern agronomy, plant breeding, agrochemicals such as pesticides and fertilizers, and technological improvements have sharply increased yields from cultivation, but at the same time have caused widespread ecological damage and negative human health effects. Selective breeding and modern practices in animal husbandry have similarly increased the output of meat, but have raised concerns about animal welfare and the health effects of the antibiotics, growth hormones, and other chemicals commonly used in industrial meat production. Genetically Modified Organisms are an increasing component of agriculture today, although they are banned in several countries. Another controversial issue is 'water management'; an increasingly global issue fostering debate. Significant degradation of land and water resources, including the depletion of aquifers, has been observed in recent decades, and the effects of global warming on agriculture and of agriculture on global warming are still not fully understood.

The agricultural world of today is at a cross roads. Over one third of the worlds workers are employed in agriculture, second only to the services sector, but its future is uncertain. A constantly growing world population is necessitating more and more land being utilised for growth of food stuffs, but also the burgeoning mechanised methods of food cultivation and harvesting means that many farming jobs are becoming redundant. Quite how the sector will respond to these challenges remains to be seen.

George R. I

FOREWORD

MOST people who are interested in any phase of English rural life and work are familiar with the Royal Show—with its thousands of choicely-bred animals, its array of the most modern farm machinery, its scientific and educational exhibits, and its crowds of country folk. Since it was instituted in 1839 it has been the greatest of our annual agricultural gatherings.

Fewer people are aware how much else the Royal Agricultural Society has done, and continues to do, in less spectacular ways, for the advancement of English farming and rural industry. It has, for instance, played no small part in the long fight against animal disease ; its implement trials have encouraged inventors and have otherwise helped towards the progressive improvement of the farmer's tools ; it has done a good deal for the promotion of agricultural research and education ; and it has, in nearly a hundred thousand pages of publications, spread abroad a vast amount of agricultural knowledge.

This book tells the story of the first hundred years of the Society's endeavour—of its mistakes and failures as well as of its wise decisions and solid achievements. It necessarily deals with the developments and the varying fortunes of the industry as a whole as well as with the Society's own affairs. It tells, too, of many men who have dedicated great abilities to the cause of farming progress, and whose lives should prove an inspiration to us, and to those who will follow on.

Athlone .

DEPUTY PRESIDENT 1939

CONTENTS

LIST OF ILLUSTRATIONS

LIST OF ILLUSTRATIONS

LIST OF ILLUSTRATIONS

CHAPTER I

ENGLISH AGRICULTURE IN 1838

THE *English Agricultural Society* was founded in 1838 and two years later was incorporated by Royal Charter as *The Royal Agricultural Society of England.* Queen Victoria became its first patron.

For a good many years before this date there had been no National Society or Institution for the promotion of English farming. The old Board of Agriculture had indeed, in its time, done a great deal to this end, more particularly so long as Sir John Sinclair and Arthur Young were giving their united energies to its work. But Governments and Ministers had grown tired of Sinclair's "intolerable energy," and in 1813 he had been displaced from the Presidency. Arthur Young had lost his eyesight in 1811 and died in 1820. There was nobody, apparently, to give the Board its old drive ; after Waterloo it gradually languished and at last, in 1822, it was dissolved. The Government Department which is now the Ministry of Agriculture came into existence only in 1889. Thus, for the first fifty years of its existence, the new Society was to stand alone. It found a great task waiting to its hand.

If we are to appreciate the magnitude and the nature of the task that lay before the founders, we must first glance at the state of England, and especially of English farming, at the time of the Society's foundation.

During the preceding seventy or eighty years the population of England had been growing with unprecedented speed. In 1760, when George III came to the throne, the population is supposed to have been about 7 millions ; in 1801, when the first census was taken, it was 10½ millions ; in 1838 it exceeded 16 millions. The increase in numbers occurred mainly in the towns—in fact, the problem of rural depopulation had already arisen in many districts. As yet only a little food was being imported ; Britain had indeed ceased, by about 1800, to be fully self-sufficient in the matter of food, but fifty years later it could still be estimated that average imports of wheat amounted to only one-eighth of the country's consumption. The problem of feeding the growing number of people was becoming ever more difficult of solution. Year by year the town-dwellers followed anxiously the progress of the farmer's crops, knowing only too well that a bad harvest must inevitably cause a shortage of bread.

Again, the Industrial Revolution had caused great changes in the distribution of the country's population. Before 1760 the density of population had been largely determined by the fertility of the land. London was the only large consumer's market. Elsewhere, and in the main, people lived where their food was grown. But other considerations determined the situation of the new factories, mines, ironworks and shipyards. Hence many farmers had to turn their eyes to more distant markets, and improved means of transport became an urgent need.

Telford and Macadam had, however, been at work for many years before 1838, and good main roads already connected all the important towns. The navigations of many rivers had been improved, and long stretches of canals had been built. Bulky produce could already be carried for considerable distances at a moderate cost, though the speed of movement was still slow. Even the new and greatly accelerated mail-coach service, instituted about the time of the Society's foundation, took 22 hours between Liverpool and London, and 42½ hours between London and Edinburgh. Perishable foods, such as milk and vegetables, could obviously not be carried far ; market gardens clustered close round the skirts of the towns, and milk for town supply was produced in town dairies, the cattle fodder being brought in from nearby farms.

Already, too, steam power was promising to revolutionise transport, as it had already revolutionised many manufactures. Twenty years earlier the first steamships had crept out to sea ; in 1829 members of the public who proposed to travel between London and Edinburgh were invited to compare the comfort of a sea passage in a new steamer, providing accommodation for a hundred passengers, with the discomfort and strain of the journey by road ; by 1838 there was already a regular steamship trade in fat cattle from Aberdeen to London, the numbers rising to several hundreds a week at the height of the Spring season.

The day of railways, too, was dawning ; the Liverpool and Manchester line had been running for eight years, and the next major development, the opening of the London and Birmingham Railway, occurred in the year that our Society was formed.

While all these changes were afoot, farming had been undergoing its own revolution. It has already been said that there had been, for two generations, a constantly increasing demand for food, with which the home supplies had failed to keep pace. In 1760 England was a food-exporting country and by 1838 had become a considerable importer. Yet the output of English soil must have increased by more than half in the period. This notable achievement did not depend upon any spectacular scientific discovery or mechanical invention. It is true that the stationary steam-engine, the threshing machine and the grain drill were all available by 1838, and that notable improvements had already been made in the plough ; but none of these things had any profound effect upon farming processes. The chemist had barely begun to throw any light upon the problems of plant and animal nutrition. So far, then, the progress of farming had, in the main, been due to the farmer's own efforts, encouraged by a friendly Government and by a hungry people.

It is possible to distinguish three chief means by which the remarkable increase in production had been attained.

Firstly, a great deal of land, formerly woodland, marsh, waste or poor sheep pasture, had been brought under the plough. The most striking examples were to be seen on the large areas of poor light land such as West Norfolk (where the new methods were first worked out), Lincoln Heath, the Wolds of East Yorkshire and the Moorlands of Northumberland. On all these broad stretches of country the problem of reclamation was essentially the same ; the soils were mostly well enough drained by nature, but some were very acid, and nearly all were grossly deficient in phosphate. The farmers who set about the task of improvement would not, indeed, have stated their problem in these terms ; but they had discovered that a mass application of lime, followed by a mass application of broken bones, would make the land fit to grow both turnips and clover ; and that, when these crops had been consumed upon the land, it became capable of producing profitable crops of wheat and barley. In 1836 the Select Committee of the House of Lords stated that Lincoln Heath, " until 35 years ago a rabbit warren," was yielding from 32 to 36 bushels of wheat to the acre. In other places there had been improvements of other kinds. The dairy pastures of Cheshire had been greatly benefited by drainage and marling, and astonishingly improved by the use of bones. The black fen soils, up till 1825, had grown little grain but oats, but the Select Committee reported that " now they are growing wheat equal to that of any part of England." Here the key to improvement had been the admixture of subsoil clay with the surface layer of light and fluffy peat.

The second main line of progress had been the replacement of the system of land tenure, and of farming, that had formerly prevailed in the great central block of England, stretching from Durham to the South Coast and from the Severn Valley to the borders of East Anglia and Kent. In 1760 the old open-field system was still the common one in this area.

By 1838 only a few parishes remained unenclosed. The great arable fields with their multitude of small strips, the village commons and the lot-meadows had almost everywhere given place to compact farms, each owned or tenanted by an individual farmer. In the process of enclosure, or as a consequence of it, many of the small farms had disappeared. Their former owners had migrated to the towns or become wage-earning workers on the land. The whole affair had resulted in a good deal of hardship and social injustice, but it had undoubtedly led to increased efficiency in the production of food, and to a considerably increased output. We must indeed remember that the Agricultural Revolution had meant different things in different places. It was an easy matter to take a stretch of the Yorkshire Wolds and carve out the almost uninhabited land into square fields and large farms, planned according to the needs of the new times. It was almost equally easy, given an Enclosure Act, to deal similarly with a Midland open-field parish. But it was otherwise in western districts —Devon and Cornwall, the Welsh Border Counties, Cheshire and Lancashire—where small enclosed farms had existed for centuries. Moreover, the inducement to large-scale capitalist farming was in these places much less strong ; corn-growing and sheep-farming, or a combination of the two, were the enterprises that attracted the new business farmer. Thus the small men in the West were left to carry on, and to adapt themselves gradually to the new order.

Apart from the spectacular changes resulting from land reclamation and enclosure, there had been a widespread and all-round improvement in the level of farming. The introduction of new crops had meant a good deal. The potato was scarcely known as a field crop in 1760, and was already important in 1838, especially in the North-West. The cultivation of turnips and swedes, of clover and " artificial " grasses, had raised the general level of fertility, so that yields of corn had risen considerably. As already indicated, lime and bones had also played an important part in raising fertility. Liming, of course, had been practised for centuries ; the value of bones was already known, here and there, by the latter half of the eighteenth century ; bones were used, for example, by farmers in the neighbourhood of Sheffield, where bone-dust and scrap were waste materials from the cutlery manufacture. Their widespread use began about 1820. Imports in 1823 were valued at only some £14,000, but the figure had risen to more than a quarter of a million pounds by 1837.

A survey of English farming in 1838 would have shown that the lighter lands, all along the eastern side of England, were already being skilfully and highly farmed. In East Anglia, on the Heath and Wolds of

Lincolnshire and again on the Yorkshire Wolds, the Norfolk four-course rotation had become standard. Here and there, indeed, clover sickness, turnip "fly" and "finger and toe" were already causing trouble, but, on the whole, the new system was proving a great success. In Northumberland the almost universal cropping scheme was a five course. Starting with oats after ley, there followed turnips, then barley or spring wheat ; with the last were sown the grass and clover seeds for a two-year ley. The seeds mixture was generally composed of perennial rye-grass, red and white clovers with a little trefoil, but might include timothy or cocksfoot or other grasses.

No corresponding development in rotational systems had occurred on the heavier arable soils. Turnips were not an attractive crop on clay land, and there was thus no substitute for the old bare fallow. Moreover, the primary need of much of the heavy land was not manure, but a more adequate system of drainage. The only system on the greater part of the English clays was the old one of ridge-and-furrow. In Essex the mole plough was already in use and there was also an ancient practice of brush draining—i.e. of making covered drains in which the channel was formed of brushwood, or sometimes of straw ; but the work was costly in labour in relation to the comparatively short life of the drains. Smith of Deanston had published his *Remarks on Thorough Draining and Deep Ploughing* in 1831, and had shown how to lay out a system of covered drains ; but his system depended on the use of vast quantities of stones, which were rarely available in the areas where drainage was most needed. A satisfactory tile-making machine was yet to be invented.

The tools and implements of cultivation were still comparatively few and simple, and often ill made. Two-horse iron ploughs were used by the larger and more progressive farmers, but a good many of the local models were heavy, clumsy and inefficient. The common type of harrow was wood-framed, square or rhomboidal in shape, with, usually, iron teeth. The grubber (cultivator or scarifier) was a recent invention, the earliest type having been patented by Finlaison in 1820. A cast-iron flat roller was the only alternative to the old wooden and stone models. Corn and root drills, and broadcast machines for grass seeds, were in common use on the larger farms. A horse-rake was a recent addition to the list. The factory-scale manufacture of standard implements—ploughs, harrows, drills, etc.—had already begun.

Patrick Bell's reaper had been invented in 1828, but its manufacture upon a commercial scale had never been taken up. Eighteen more harvests were to be reaped by hand before an alternative means was to be seriously considered. Meikle's threshing-machine had been invented in 1785 and it, or some other type, driven by a stationary engine, by water or by horse-power, had long ceased to be an object of wonder in the North and North-East. In the South country, and especially in the heavy-land districts, the flail had lingered much later ; here there was little winter work except threshing for the large number of men who could be employed in summer, and there was little to be gained by buying a machine and standing off the flail-men. Before 1830, however, the machines were being widely introduced. Their introduction was not the sole cause of the riots and rick-burnings organised by the mysterious " Captain Swing " in 1830-31 ; the fact was rather that the labourer regarded the machines as likely to add the last straw to his already almost intolerable burden of poverty.

The extent of arable land was of course much greater than it is at the present day. Corn crops, and especially wheat, were of outstanding importance. No official statistics of acreage were collected before 1866, but various estimates were made by different writers. Probably the most reliable are those published by McCulloch in his *Statistical Account of the British Empire*, and his figures, taken from the edition of 1846, are compared below with those of the present day :

ACREAGE OF CHIEF ARABLE CROPS IN ENGLAND AND WALES

(*Thousands of Acres*)

	1846	1938
Wheat .	3,800	1,829
Barley .	1,500	886
Oats and Rye	2,500	1,410
Beans and Peas	500	269
Potatoes, Turnips, Rape, etc.	2,000	1,552
Clover (and Rotation Grasses)	1,300	1,900
Bare Fallow .	1,500	351
Hops .	50	18
Total Arable Area (excluding fruit and market garden crops) .	13,150	8,215

As regards wheat, it will be seen that this occupied nearly half the total area devoted to cereals, and well over a quarter of all the arable land.

None of the very numerous varieties of those days—Kentish Yellow, Burwell Red, Golden Ear, Velvet-Chaff, etc.—seem to have survived, except only the old Cone or Rivet. Many of them were of very local distribution. Probably one sort went by different names in different places, and many of the so-called varieties were certainly mixtures of strains. Various estimates of average yields might be quoted. Professor

-6-

Low,[1] who is a reliable authority, says : " The produce of this crop varies greatly with the season, the nature of the soil and the mode of cultivation. A fair good crop may be held to be about thirty bushels per acre. The average produce of England does not perhaps exceed twenty-two bushels nor that of Scotland twenty-five." These figures compare respectively with estimated averages of 31½ bushels and 38½ bushels to-day. There can be no doubt that yields varied more widely, from year to year, than they do now. A wet winter meant always a poor yield from the heavier soils, while rust, smut, and especially bunt, often caused serious loss.

Barley was grown where the land was not in sufficiently high condition, or was otherwise unsuited for wheat ; oats were grown where neither of the other cereals was likely to succeed. Rye, once an important crop on the lighter soils such as the sands of Nottingham, had ceased to be widely grown. The Scotch Potato oat, which still survives, was the standard variety in the North, while Chevallier barley, originally selected in 1823, was becoming widely known. Otherwise the varieties mentioned in the books of the time have all but disappeared.

Turnips and swedes were by now widely cultivated and were given great care and attention. The turnip land received a heavy dressing of dung or of bones, or perhaps more commonly moderate quantities of both. Broadcast sowing had given way to drilling, and the crop was singled, and hoed both by horse and by hand. In the North, crops of twenty-five tons of common turnips, or twenty of sweeds, were not regarded as extraordinary. Mangolds had been introduced about 1800, but the acreage grown was still very small.

Lancashire had been the first county to take up the field culture of the potato, and the crop had long been largely cultivated there. The County Survey of 1795 already speaks of yields of eight to twelve tons per acre, discusses the advantages of using Scotch seed, and describes a new process, worked out by Mr. Blundell of Ormskirk, of chitting or sprouting the sets before planting. Blight was, of course, as yet unknown, but " The Curl " (Leaf Roll and other virus diseases) was already recognised as a serious trouble, and some methods of prevention, such as the early lifting of crops intended for seed and the use, for seed, of potatoes grown on high-lying farms, were well known. There must have been some considerable cultivation in the Vale of York, for Brown, in his *Agriculture of the West Riding* (1799), says : " Large quantities of potatoes are sent by water carriage, from Selby and other ports of the River Ouse, to London market." On the other hand, Arthur Young in his *General View of the Agriculture of Lincoln*

[1] *Elements of Practical Agriculture*, 2nd Edn., 1838.

(2nd Edition, 1813) places the potato among " Crops not commonly culti-vated," though he mentions one particular farmer, near Spalding, who " always had 12 to 15 and once more than 20 acres." Large-scale culti-vation in Lincolnshire started only about 1880. It is difficult even to guess at the total potato acreage cultivated in 1838. By 1867 it exceeded 300,000 acres, or substantially more than half that of the present day ; but in the interval there had been a large increase.

The list of arable crops included, besides beans, peas, hops, flax, cabbage, carrots, etc., a number of minor ones which have now gone out of cultivation. Among them were hemp, woad, teasles, and buckwheat.

As regards the species of pasture plants, broad red and late-flowering red clovers, Dutch white clover and trefoil were all well known. Alsike clover had just been introduced (1834) and thus, except for wild white clover, the list of clovers was as we know it to-day. Perennial rye-grass was by much the most important of the sown grasses, but most authors of the period include foxtail, timothy, meadow fescue, rough-stalked meadow-grass and cocksfoot as worthy of a place in mixtures. Italian rye-grass was a promising novelty, having been introduced in 1833.

Turning to the country's animal husbandry, the sheep population was probably not notably different from that of to-day, whereas the numbers of cattle were certainly less than half. The earliest breed of cattle to be im-proved, the Longhorn, was still to be found in Lancashire, and a few pedi-gree breeders were still able to find a market for their bulls, mainly in Ire-land. But the breed had, twenty or thirty years earlier, been driven from its leading position by the Shorthorn. Of the famous improvers of the latter, the brothers Colling had lived and died, Thomas Bates, John and Richard Booth were at the top of the tree, and Amos Cruickshank, who was to make history half a century later, had just become tenant of the obscure farm of Sittyton in Aberdeenshire. In 1838 the Shorthorn, and the related Holderness and Lincoln cattle, were reckoned the best of the native milch breeds.

The Hereford and Devon had already been specialised as beef breeds, and both were very highly esteemed by graziers in the fattening areas and by London butchers. The South Devon and the Sussex were of no more than local importance ; the Red Poll breed had been founded, but was still unknown to most of the writers of the time ; one of its parents, the Polled Suffolk Dun, still survived. Large numbers of Scotch cattle (by this time mainly Galloways) and Welsh " Runts " were brought every year to England to be fattened. There was also a considerable import of Irish stores—some Longhorns, some Shorthorn crosses and a good many nondescripts. Channel Islands cattle, still generally known as Alderneys but in fact mostly Jerseys, had been imported in quite considerable numbers during the previous twenty or thirty years.

Cattle of the better and earlier-maturing breeds were, by the time of which we are writing, quite commonly fattened at three or three-and-a-half years old. The common practice in winter feeding was to give a moderate ration of hay or straw, some two or three pounds of linseed cake (the only concentrate commonly used) and an unlimited supply of turnips. Professor Low reckoned that " An ox of 50 or 60 stone will consume about a ton of turnips in the week or about an acre in 24 weeks," and adds that " If he thrive well he will gain in weight 14 lb. or more in the week."

The production of milk for the liquid market was confined to town dairies and farms lying within a ten- or twelve-mile radius of the consuming centres. The city cows were fed in summer mainly on green grass or fresh forage, brought in daily by wagon. In winter they had hay, turnips or other roots or cabbage, with generally some milling offal and beans and, in many cases, distiller's or brewer's grain. In rural dairying districts, such as the Vale of Aylesbury, the Cheshire Plain and the Somerset, Gloucester and Wiltshire Vales, the great bulk of the milk was converted on the farm into butter or cheese. Cheshire, Cheddar, Single and Double Gloucester, Stilton and Wiltshire were the most important market types, though there were many other local varieties. In Suffolk the practice was to make both butter and skim-milk cheese. The colouring of dairy products was a common practice, and for this purpose annatto had already replaced the older materials such as carrot-juice and marigold flowers.

It is possible to find individual records of very high milk yields, but the average was only moderate. The common reckoning was that a useful cow should produce 150 lb. of butter or 4 cwt. of cheese in a year, both of which figures would indicate a yield of about 450 gallons of milk.

In the sheep world, the attempts, begun in the late eighteenth century, to acclimatise the Merino (in order that England might compete in the market for fine wool) had been almost completely abandoned, and all breeders were aiming primarily at mutton qualities. The writers of the time mostly enumerate thirty or forty breeds, but many of these, such as the Norfolk Horn, the Teeswater and the Cannock Chase, were soon afterwards to be crossed out of existence. Two breeds were of outstanding importance—the New Leicester and the Southdown. These had already been highly improved, the one by Bakewell and the other by Ellman. Among other lowland sorts, the Cotswold, Lincoln, Dorset Horn, Ryeland, Romney Marsh, the Devon and the South Devon were recognised as distinct breeds, and were in process of improvement. The Suffolk and Oxford were already in process of formation, the Hampshire, the Shropshire and the Wensleydale not yet. In the North, the Cheviot was being kept in preference to the Blackface wherever possible, its finer wool being still a considerable asset. Different breeds of the Blackface Mountain group were not separately distinguished. The Welsh Mountain and the Herdwick were recognised as district breeds, but the other sheep of Wales seem to have been rather mixed and nondescript.

The improvement of the sheep themselves, and the introduction of turnips, had led to a considerable speeding-up of the process of mutton production, but fat lambs and even fat tegs rarely appeared at markets. Some house-fed lamb was produced in Dorset and sold in London. The bulk of the output from lowland flocks reached the butcher at some age between fifteen and twenty-seven months. Thus in summer the supply consisted of both shearling and two-shear sheep. Leicester wethers commonly reached a carcass weight of 60 to 80 lb. as grass-fed shearlings, 80–100 lb. as winter-fed shearlings and 100–120 lb. at two shear. Store sheep, in winter, rarely received any food but turnips and hay, but fattening flocks usually had a little corn, mixed with linseed cake or brewery or distillery grains. Mountain sheep rarely went to the block younger than two and a half, and many were three and a half years old. Summer folding on vetches, sainfoin, rape, etc., was a common practice in the chalk and limestone districts of the South and East.

Of the three draught breeds of horses the Old English Black Horse (The Shire) was still rather variable in type, in keeping with the wide range of country which it occupied and the variety of land on which it was bred. The Clydesdale and the Suffolk were more uniform and more definite breeds. The Cleveland Bay had spread to some extent beyond its native district, and was in large request for stage and private coaches.

The breeds of pigs mentioned in the writings of the time are the Chinese, Rudgwick, Berkshire, Essex, Shropshire, Woburn and Dishley, with sometimes the Yorkshire and others. The Berkshire was generally regarded as the most highly improved, but at this stage in its evolution it was " white or sandy in colour, with black spots and drooping ears."

The feeding-stuffs for pigs included separated milk and whey, oats, barley, beans and peas, milling offals, carrots and potatoes. Stores and in-pig sows were commonly folded on clover, vetches, etc., and in woodland districts there was still some fattening on acorns and beech-mast. There was some market for fat sucking-pigs. Porkers were commonly sold at six or eight months of age, weighing perhaps six score. Baconers were commonly kept till ten to twelve months, and occasionally till eighteen months, before slaughter. Bacon was cured only in winter, and generally on the farm.

Fowls were kept, of course, on almost every farm, but there seem to have been no large-scale or highly specialised poultry farming except the fattening of Surrey chicken for the London market. This was already an old industry, and the cramming of the birds was a regular practice. Of the many breeds, the white Dorking and the Poland were those most highly esteemed for utility purposes, while Game birds were quite extensively bred for sport. Two passages from Low's *Practical Agriculture* throw an interesting light on the changes that have been brought about, in the bird itself, during the past century. In the one he quotes, as an illustration of the high level of productivity that had then been achieved, the egg record of a pen of five Poland hens. In eleven months the five birds had laid 503 eggs with an average weight of $1\frac{5}{16}$ oz. On the other hand, he says, " Hens are at their prime at three years old, and decline after the age of five."

In parts of Buckinghamshire and Berkshire there was an extensive cottage industry in the rearing of ducklings for London. Norfolk, again, had long been famous for its turkeys. Geese had declined in numbers with the disappearance of so many of the commons : dovecotes were in many places falling into decay, and those that remained in use were universally regarded as unwanted survivals.

On a superficial consideration of the relative levels of prices and agricultural wages which prevailed, we may find it hard to believe that the farmer, or at least the larger tenant farmer, could have been anything but prosperous. For the ten-year period ending in 1838 the average price of wheat was 56s. 3d. per quarter, while farm workers' weekly wages averaged about 10s. a week in the South-East, about 9s. in the Midlands and South-West, and about 12s. in the North. The Dorset labourer was the lowest paid at 7s. 6d., and the Cheshire wage the highest at 13s. It is true that the actual cash wage always puts the worst complexion on the labourer's position, for he has always had the advantage of a cottage at a nominal rent, and other perquisites of some value. In most districts, however, the farmer could have paid five or six men their weekly wages with the price of a quarter of wheat. But this is too simple a calculation to give a true or complete picture of the position. For one thing, while wheat then stood at a high price, according to our modern standards of value, most other kinds of agricultural produce were fairly cheap. In the week in May 1838, when the founders of the Society met at the Freemasons' Tavern in London, wheat was worth 62s. a quarter, but barley was 29s. 6d. and oats

23s. 3d., while both beef and mutton were selling on Smithfield Market at 5½d. to 6½d. per pound. In the same year Cheshire cheese stood at about 70s. per cwt., which would imply a gross return of about 7½d. per gallon for milk ; Southdown wool was worth about 1s. 6d. a pound and Blackface mountain wool about 6d. Five years earlier, after a fairly abundant harvest, the London grain prices were about 45s. for wheat, 28s. for barley and 19s. for oats. The fact is, then, that the general price level of agricultural commodities was not nearly so high as the price of wheat would suggest.

We must also remember that the largest item in the farmers' expenditure was not wages but rent ; thus Professor Low gives very full estimates of the capital, income and expenditure for a typical 500-acre farm, in which the annual wage bill is put at £403 and the rent and burdens at £1,007. The fact is that, from 1814 till about 1836, farming had been suffering serious depression. " The attention of Parliament was continually called to the distress of the landed interests. Petitions covered the table of the House of Commons . . . Rural conditions were deplorable." [1] Government Select Committees on the state of agriculture were appointed in 1820, 1821, 1822, 1833 and 1836, and only the last of these could discern any definite sign of betterment.

The causes of distress were partly the same, in 1815–36, as in 1920–36. In both cases there had been a major war, and in both, war conditions had led to inflated prices ; in both, again, heavy taxation was required to meet the interest upon war debt. In both, a return to the gold standard had led to a steep fall in commodity prices. In neither case was the farmer able to reduce proportionately, or quickly enough, the major item in his outgoings. In the earlier period that item was rent, in the latter wages. The depth of depression had been reached in 1830–31, when the weather joined the other forces arrayed against the farmer. In that year, among other disasters, two million sheep had died of the rot.

Two recent Acts of Parliament had done something to alleviate the position. The Poor Law Amendment Act of 1834 had put an end to a demoralising system of Relief and had considerably lightened the heavy burden of poor rates. The Tithe Commutation Act of 1836 had not, indeed, as time was to show, settled the thorny problem of tithe for good and all ; but it had removed a good many hardships, had done away with the irritating and wasteful practice of collecting tithe in kind, and had removed an obstacle to agricultural progress ; the improving farmer could look forward to reaping the full reward of his efforts. Moreover, by 1836

[1] Ernle : *English Farming, Past and Present.*

prices had become comparatively stable, and rents had been adjusted to a corresponding level.

One dark cloud, however, hung on the horizon. The serious and persistent campaign against the Corn Laws had well begun. The Anti-Corn-Law League was to be formed only six months after the Royal Agricultural Society.

On the other hand, the founders of the Society saw another hope— that Science would at last come to the farmer's aid. Lawes had begun his experiments at Rothamsted in 1834, and had, the next year, secured the co-operation of Gilbert. Probably Liebig was already turning his master brain to the mysteries of soil fertility and plant growth. There were already reports of the remarkable effects of a new manure called guano. Shireff, in Scotland, was busy selecting cereals on a systematic and almost scientific plan. The nature of plant disease was beginning to be understood. The first actual trial of steam power, for tillage, had taken place in 1836. Scientific discovery and invention seemed likely to speed up the progress of the whole industry. It was this hope which inspired Earl Spencer to choose, as the Society's motto, the words *Practice with Science.*

CHAPTER II

THE FOUNDATION OF THE SOCIETY AND ITS FOUNDERS

IT may seem strange that a National English Agricultural Society should have been formed no earlier than a century ago. The Bath and West Society dates from 1777, the Highland Society from 1784 and the Smithfield Club from 1798. The explanation probably is that the England of earlier times was too large a unit to be embraced in a single agricultural association. In those days farmers in Northumberland, Cheshire, Norfolk and Kent knew little of each other's doings and had few interests in common. A national Fat-stock Show in London was indeed a possibility because London was by far the most important meat market, and already drew supplies from the most distant parts of the country ; but a representative exhibition of English live stock was hardly to be thought of. Thomas Bates's undertaking of bringing a group of Shorthorns from Darlington to the first Royal Show at Oxford [1] was one that few breeders would have cared to attempt. By 1839 transport facilities had improved to some extent, but the holding of a regular annual national show, which was to be the largest of the Society's various undertakings, would have been hardly possible without the system of railways which, in 1838, was still in a very early stage of development.

The movement for the formation of the Society was started by John Charles, 3rd Earl Spencer, and he, appropriately, became its first President.

Earl Spencer (Lord Althorp) was born in 1782 and entered Parliament as member for Okehampton in 1804. In the earlier years of his parliamentary career he made no great impression on the House of Commons, and was not thought of as a coming man. Indeed, he seemed to lack most of the qualities that make for success in politics, being shy, almost tongue-tied and without personal ambition. Gradually, however, his solid ability, his industry and, above all, his absolute honesty won the general respect of the House and his halting speeches began to be heard with ever-growing respect. In 1827 he was made leader of the Whig

[1] See below, p. 20.

party, then in opposition ; three years later, when his party came into power, he became Chancellor of the Exchequer and leader of the House of Commons. But he was at heart a country squire, and it was nothing but his strong sense of public duty that kept him in political work. While still a busy politician he was master of the Pytchley Hunt, and was absorbed in trying to evolve a new type of hound ; but far before budgets, and even some way before hounds and hunting, came the herd of Shorthorn Cattle which he kept on his wife's estate at Wiseton in Northamptonshire.

In 1834 his father died and he succeeded to the family title. He was urged by his friends to remain in politics, but he had had enough. Moreover, he found his estates heavily burdened with debt, and felt that they required his undivided attention. His sole extravagance from this time onwards, as he once said, was farming.

In 1825 Althorp had been elected President of the Smithfield Club, which was then in such financial difficulties that it seemed destined for an early demise. Very largely by his work its finances were put upon a sound basis and it took on a new lease of life. On one occasion, on the morning of a show, when something had gone wrong with the arrangements, he was found by an early visitor working among the stalls in his shirt-sleeves.

The annual dinner of the Smithfield Club, held in the Freemasons' Tavern in London in December 1837, was the occasion chosen by Earl Spencer for suggesting the formation of a new society for the promotion of agriculture. In proposing, as President, the Success of the Smithfield Club, he remarked that, while the improvement of cattle was a very worthy object, there were many other sections of the farming industry which were equally capable of improvement ; an association was needed which would cover the whole field. " The application of science to practice was not as yet made by the English farmer. . . . If a society were established for agricultural purposes exclusively, he hesitated not to say that it would be productive of the most essential benefits to the British farmer. . . . There was one point, however, which he must strongly impress upon them, namely, that there could be no prospect of their obtaining any useful results unless politics . . . were scrupulously avoided at their meetings."

The reason for Spencer's insistence that politics should be barred is obvious if we remember the circumstances of the time. Controversy about the Corn Laws was rising, and it seemed likely that this would be long and bitter. Cobden's Anti-Corn-Law League was to be formed in

1839. Many of the most progressive landowners were Members of Parliament, and they were by no means all on one side. The Society might have split from top to bottom on this or some other political controversy, with the most unfortunate results to its other activities.

Probably with the idea of emphasising the absence of any political objective, Spencer had arranged with the Duke of Richmond, a tory, that he should be the second speaker in support of the proposal. The Duke spoke of the good work of the Highland Society, and saw no reason why the farmers of England should not follow so excellent an example. Among the many others who supported the proposal was Henry Handley, M.P. for Lincolnshire, who was to be prominent in the Society's councils for many years.

In the following month (January 1838) Handley addressed a long letter to Spencer setting out the various services that the proposed Society might render to agriculture, and giving his views about its constitution and organisation. This letter was printed, and copies were widely circulated. Some months were allowed for consideration and private discussion and then, early in May, the following advertisement was inserted in the *Morning Herald* :

THE NEW AGRICULTURAL SOCIETY

The undermentioned noblemen and gentlemen, having observed the great advantages which the cultivation of the soil in Scotland has derived from the establishment and exertions of the *Highland Society*, and thinking that the management of land in England and Wales, both in the cultivation of the soil and in the care of woods and plantations, is capable of great improvement by the exertions of a similar Society, request that those who are inclined to concur with them in this opinion will *meet* them *To-morrow*, the 9th of May, at the *Freemasons' Tavern*, at one o'clock, to consider the means by which such a *Society* may be *established*, and of the regulations by which it shall be governed.

It is suggested that the Society shall be called " *The English Agricultural Society*," and that it shall be one of its fundamental laws that no question be discussed at any of the meetings which shall refer to any matter to be brought forward or pending in either of the Houses of Parliament.

It is also suggested that the Society shall consist of two classes of subscribers —the one to be called Governors, subscribing annually 5*l*. ; the other Members, subscribing annually 1*l*.—either the one or the other to be permitted to become Governors or Members for their lives by the payment, in one sum, of the amount of ten annual subscriptions.

Gentlemen wishing to subscribe to this institution will be good enough to send their names and addresses to the editors either of the *Mark Lane Express* or of *Bell's*

Weekly Messenger, specifying whether they wish to become Governors or Members, and whether they wish to subscribe annually or for life.

DUKE OF RICHMOND	R. A. CHRISTOPHER, ESQ., M.P.
DUKE OF WELLINGTON	JOHN BOWES, ESQ., M.P.
EARL FITZWILLIAM	H. BLANCHARD, ESQ.
EARL SPENCER	W. T. COPELAND, ESQ., M.P.
EARL OF CHICHESTER	J. W. CHILDERS, ESQ., M.P.
EARL OF RIPON	WILBRAHAM EGERTON, ESQ.
EARL STRADBROKE	RALPH ETWALL, ESQ., M.P.
MARQUIS OF EXETER	H. HANDLEY, ESQ., M.P.
LORD PORTMAN	C. SHAW LEFEVRE, ESQ., M.P.
LORD WORSLEY, M.P.	WALTER LONG, ESQ., M.P.
HON. ROBT. CLIVE, M.P.	WM. MILES, ESQ., M.P.
HON. BINGHAM BARING, M.P.	JOS. NEELD, ESQ., M.P.
HON. C. C. CAVENDISH, M.P.	E. W. W. PENDARVES, ESQ., M.P.
SIR ROBT. PEEL, BART., M.P.	PHILIP PUSEY, ESQ., M.P.
SIR JAMES GRAHAM, BART., M.P.	E. A. SANFORD, ESQ., M.P.
SIR FRANCIS LAWLEY, BART., M.P.	R. A. SLANEY, ESQ., M.P.
SIR WATKIN OWEN PELL, BART.	J. A. SMITH, ESQ., M.P.
LIEUT.-GEN. SIR E. KERRISON, BART., M.P.	R. G. TOWNLEY, ESQ., M.P.
	W. WHITBREAD, ESQ.
EDWD. BULLER, ESQ., M.P.	HENRY WILSON, ESQ.

The list of names contains those of many men prominent in public affairs, or in agriculture, or both. Besides Spencer himself, the Duke of Richmond and Handley, the men who were to play the most important parts in the direction of the Society were Shaw Lefevre and Philip Pusey. The former, indeed, was later elected Speaker of the Commons and was thus obliged to give up active work for the ' Royal.' Pusey, on the other hand, devoted himself very largely, until his death in 1855, to the Society's business, and especially to the direction and editing of its *Journal*.

On the following day (May 9, 1838) the meeting-room at the *Freemasons* was " crowded to excess." Spencer was voted to the chair, and the resolution for the formation of the Society was moved by the Duke of Richmond and seconded by Handley. It was then found (though evidently Spencer was forewarned) that the meeting contained a small but determined opposition. A meeting of farmers had been held the previous evening in London and had passed a resolution declaring that the proposed Society was " delusive in principle, having a tendency to mislead the farmers of England and betray the interests of English Agriculture, in so far as the principle of legislative protection to agriculture is disavowed in the fundamental resolutions." Two members of the opposing party spoke at some length on the " delusiveness " of the project, and their contention is quite easily to be understood. They had conceived the idea of a society like our present-day National Farmers' Union, whose main purpose would have been to take action, on behalf of farmers, in political affairs. There was some considerable commotion, but Spencer pointed out that he and the other signatories of the advertisement had called a meeting of persons who were in favour of forming a Society of a non-political kind, and that the opposition was therefore out of order. The protesting minority then withdrew, held a separate meeting, and decided to form a separate association. This took the title of " The Farmers' Central Society of Great Britain and Ireland." Its objects were defined as " The protection and encouragement of agriculture in all its branches, without reference to party political feeling." It was evidently intended as a rival body to the ' Royal,' but it seems, in fact, to have done little except to join in the losing battle against the repeal of the Corn Laws.

The meeting, after the withdrawal of the dissentients, proceeded to pass unanimously a number of resolutions about the general constitution of the new Society, and to elect a provisional committee. The meeting also elected Spencer as the first President and Mr. William Shaw, then editor of the *Mark Lane Express*, as Secretary. The name chosen was " The English Agricultural Society," the change to the present name having been made in the Charter two years later.

The Provisional Committee met on May 12 and drew up the following list of the objects of the Society :

1. To embody such information contained in agricultural publications and in other scientific works as has been proved by practical experience to be useful to the cultivation of the soil.

2. To correspond with agricultural, horticultural, and other scientific societies, both at home and abroad, and to select from such correspondence all information which, according to the opinion of the Society, is likely to lead to practical benefit in the cultivation of the soil.

3. To repay to any occupier of land, who shall undertake at the request of the Society to try some experiment how far such information leads to useful results in practice, any loss that he may incur by so doing.

4. To encourage men of science to the improvement of agricultural implements, the construction of farm buildings and cottages, the application of chemistry to the general purposes of agriculture, the destruction of insects injurious to vegetable life, and the eradication of weeds.

5. To promote the discovery of new varieties of grain and other vegetables useful to man or for the food of domestic animals.

6. To collect information with regard to the proper management of woods, plantations, and fences, and on every subject connected with rural improvement.

7. To take measures to improve the education of those who depend upon the cultivation of the soil for their support.

8. To take measures for improving the veterinary art as applied to cattle, sheep, and pigs.

9. At the Meetings of the Society in the country, by the distribution of prizes and by other means, to encourage the best mode of farm cultivation and the breed of live-stock.

10. To promote the comfort and welfare of labourers, and to encourage the improved management of their cottages and gardens.

On June 26 the Committee recommended that the first Country Meeting should take place at Oxford on July 17, 1839, and nominated fifty persons to form a Committee of Management. The list includes the names of landowners like Shaw Lefevre, Pusey, Sir Harry Verney, John Heathcote and H. S. Thompson ; of leading farmers such as John Ellman of Glyde (the son of the famous Southdown breeder) and Fisher Hobbs of Coggeshall, Essex (already famous for his arable farming) ; of Colonel Le Couteur, the improver of wheat, and of William Youatt, the well-known veterinarian and writer on live stock.

The first General Meeting of the Society was held on the day following, when the Committee reported that the number of Governors and Members had already reached 466, and that over £2,500 had been received in subscriptions. The Committee had already decided to offer prizes for essays, and a list of eighteen subjects was published at the meeting, including The Rotation of Crops, The Analysis of Soils, The Application of Mechanical Power in Farming and The Diseases of Plants. Among suggestions emanating from members was one that the Society should acquire chemical apparatus and appoint an able chemist to analyse soils and to recommend the use of appropriate manures.

Thus in less than two months the Society seemed to be established on a firm basis, with a clear programme, a considerable bank balance and an active committee of enthusiastic and well-informed men, carefully chosen to represent the many interests of the industry and many different branches of knowledge.

At the next general meeting, held in December, Earl Spencer announced that it had been decided to publish a Journal of the Society's Proceedings. The management of the publication was placed in the hands of a special committee, and the editorial control in those of Philip Pusey, who was soon to make the publication an important force in the improvement of farming.[1]

For the following six months the Committee of Management was occupied mainly with the arrangements for the Oxford Show. There seem to have been various discussions about the Society's motto and, in the end, the choice was left to Spencer. The motto " Practice with Science " was chosen by him, and has since been retained.

The Prize List for the Oxford Meeting was published in the form of a huge placard, a copy of which is preserved in the Society's offices. It is clear that the first Country Meeting was awaited with the greatest interest and excitement. It was obviously a heavy undertaking for exhibitors to send live stock and implements to a centre where there was no railway, and it was even no small affair for an interested farmer, living at a distance of one or two hundred miles, to get himself to the Show.

Thomas Bates, for instance, brought four Shorthorns (a bull, a cow and two heifers) from Kirklevington in Tees-side. The cattle were driven from the farm to Hull, and thence went by boat to London, accompanied by their owner. They were transhipped to a barge and taken by canal to Aylesbury. There they remained one night ; the next day they were driven the ten miles to Thame, and finally, the day after, the remaining thirteen miles to Oxford.

A group of visitors from Tavistock in Devon put up one night at Exeter and another at Cheltenham, travelling part of the way by post-chaise and part by stage-coach.

The site of the Show was " Mr. Pinfold's pasture ground, Holywell " (now occupied by Mansfield College), and its seven acres seem to have provided all the space that was required for the exhibits, though the number of visitors was to prove uncomfortably large. A hand-bill of general regulations was widely circulated, and is reproduced on the opposite page.

During Sunday and Monday, the live stock and implement exhibits arrived, and by the morning of Tuesday, July 16, the arrangement of the exhibits was complete. Tuesday morning was devoted to implement trials in a field adjoining the showyard—a combined seed-and-manure drill for turnips, a subsoil plough, a scarifier and some few more. In the afternoon there was a paper-reading conference at the Town Hall. Colonel Le Couteur, of Jersey, read his prize essay on varieties of wheat, Mr. Handley discussed the comparative advantage of wheel and swing

[1] See Chapter XII.

ENGLISH AGRICULTURAL SOCIETY.

General Regulations for the Oxford Meeting, July 17, 1839.

SHOW YARD,

In Mr. Pinfold's Pasture Ground, Holywell.

No Stock can be admitted for exhibition unless the necessary Certificates, in the form prescribed, and signed by the Exhibitor in the manner directed, be delivered to the Secretary, or sent post paid, so as to reach the Society's Rooms, 5, Cavendish-square, on or before the 1st of July.

Non-Subscribers to pay Five Shillings for every head or lot of Live Stock before obtaining a ticket of permission to bring their Cattle into the Show Yard.

Persons intending to exhibit Extra Stock must give notice to the Secretary on or before the 1st of July.

No Stock will be admitted into the Show Yard before Seven o'clock in the Morning, nor later than Nine o'clock on Monday Evening, nor before Four o'clock on the following Morning, Tuesday the 16th of July; and Stock of every description must be in the Show Yard before Eight o'clock that Morning, and will remain in the charge of the Society until Seven o'clock on Wednesday Evening.

No Animal can be removed during the Show without an order obtained from the Stewards of the Show Yard.

Whenever reference is made to Weights or Measures, it is to be considered that the Imperial Weights and Measures are alone referred to.

Persons intending to exhibit Implements, Roots, Seeds, &c. must give notice of their intention to the Secretary, and furnish him with a description, on or before the 10th of July; and all such Implements, Roots, Seeds, &c. must be brought to the Show Yard on Monday the 15th of July.

Tickets of Admission to the Show Yard can be had before Twelve o'clock on Wednesday Morning the 17th of July, by exhibitors gratis; by the Public at a charge of Two Shillings and Sixpence each; and after Twelve o'clock that Morning the Public to be admitted at One Shilling for each person.

The Tickets to be had at the entrance into the Show Ground, and at the Star and Angel Hotels.

All Stock, and other Articles for Exhibition, to enter the Show Yard at the approach by Holywell Church, and Visitors to enter from the Parks, next Wadham College Gardens.

No Person to be allowed to enter the Show Ground without a ticket of admission.

Arrangements will be made for a Sale by Auction on Thursday the 18th, in the Show Yard, of such portions of Stock exhibited as Proprietors may decide to submit for Sale, of which due Notice will be given, and Catalogues prepared.

Admittance to the Yard, on the day of Sale; One Shilling each person.

All Exhibitors having Animals for Sale are requested to give notice thereof to the Secretary, 5, Cavendish Square, on or before the First of July.

DINNER,

In Queen's College Quadrangle, High Street,

On WEDNESDAY the 17th of JULY.

Books to be opened both at 5, Cavendish Square, and at the Star Hotel, Oxford, for the insertion of Names of Members of the Society desirous of engaging Tickets for the Dinner, which will be kept open until the First of July.

Tickets will be reserved for such applicants at Oxford.

LODGINGS, &c.

Registers will be kept at the Bars of the Star and Angel Hotels, Oxford, to enter particulars of Lodgings, &c. offered for the occasion; and all Persons having Rooms, Stables, or Coach-Houses to Let, or requiring such accommodation, are requested to apply to Mr. GRIFFITH, at these Hotels.

N. B. On Tuesday the 16th Trials of Agricultural Implements will take place; and the Prize Essays will be read in the Town Hall, open to all Subscribers to the Society.

Committee Room, Star Hotel, Oxford, June 15, 1839.

H. HALL. Printer, Oxford.

ploughs, the President gave an analysis of his records of gestation in cows, and there were two other contributions.

Next morning Oxford was early astir. The showyard was open from seven o'clock, and from that hour onwards, and until the late afternoon, an almost continuous stream of visitors passed through the gates. The report in *Bell's Weekly Messenger* says :

> The influx of visitors from many miles round Oxford was exceedingly great, the principal streets being completely lined with gigs, coaches and other conveyances, while the town throughout the whole day presented such a scene of bustle as was never, perhaps, before witnessed. The crowd waiting for admittance to the Show-yard was so extensive that, immediately the gates were thrown open, the rush was so tremendous that many gentlemen had their coats torn from their backs. Although 5000 tickets of admission had been printed, before ten o'clock the whole had been disposed of at 2/6 each. The consequence was that some thousands who were unable to obtain them were refused admittance.

After one o'clock the price of admission was 1*s*., and 12,000 tickets had been printed ; but these also were soon sold out, and some members of the Committee were deputed to take money at the gates. Altogether the attendance exceeded 20,000 and the total " gate " reached £1,189. The day, fortunately, was very fine, but to the discomfort of overcrowding in the showyard was added the difficulty of obtaining food, all the inns in the town being thronged with clamorous, hungry men. More than one visitor reported that he set out on his return journey in the evening in a famishing condition.

The judging of the live stock had taken place on the previous day in the absence of the public, and with the most elaborate precautions to ensure that the judges would be ignorant of the ownership of the animals. Apparently the awards were not publicly announced until the great dinner, held in the quadrangle of Queen's College, on the Wednesday afternoon.

There were separate classes, in the cattle section, for Shorthorns, Herefords and Devons, the numbers of entries being 27, 24 and 15 respectively. For each of these breeds one prize was given for (*a*) A Bull, (*b*) A Cow in Milk, (*c*) An In-calf Heifer, (*d*) A Yearling Heifer, and (*e*) A Bull Calf. The prize in the first class was £30, in the second and third £15, and in the fourth and fifth £10. A similar classification, according to age, was provided for " any other breed or cross," the animals exhibited being Longhorns, Sussex and various crosses. Two prizes were awarded for cows in milk, the winner being a nine-year-old Hereford and the second a Shorthorn which was in her fifteenth year.

IIA.—A Scene at Wiseton
Earl Spencer at Left
From a painting by Richard Ansdell

IIB.—The English Agricultural Society's Showyard at Oxford, 1839
The Duke of Richmond is on the right, listening to Henry Handley. Earl Spencer, hands in pockets, is in the background

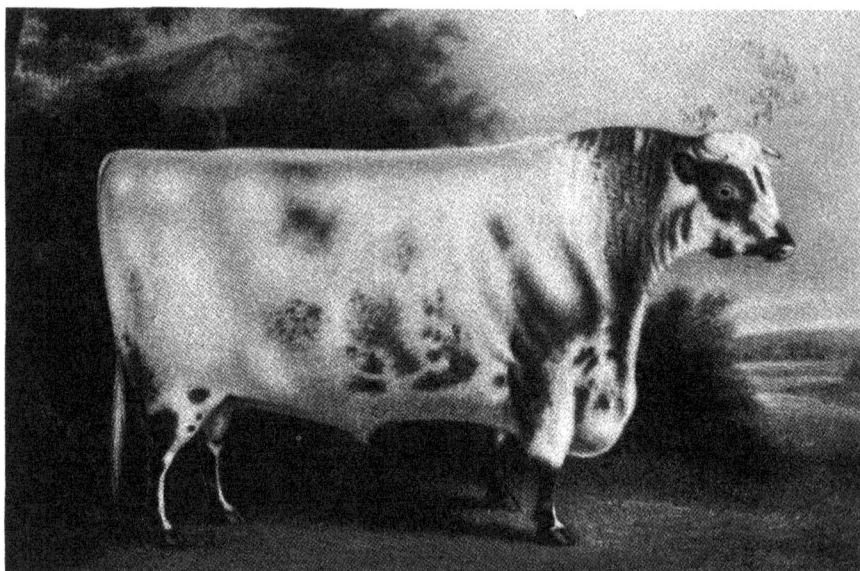

IIIa.—Thomas Bates's "Duke of Northumberland"
First Prize Shorthorn Bull, Oxford, 1839

IIIb.—Hereford Cow and Calf
From Low's "Domestic Animals of the British Islands," 1842

The cattle section was completed with a class for groups of five oxen. The winning group were Herefords and the second Devons.

Ten Cart Stallions competed for a £20 prize, and this was awarded to a Suffolk, although the breed is not specified in the award list. Six cart mares were entered, and the winner was probably a Shire. A premium of £30 had been offered for a Hunter, Carriage or Roadster sire, but the judges considered that none of the eight entries reached a sufficient standard of merit, and the award was withheld.

The four age-groups of Leicester Sheep (Shearling Ram ; Ram of other age ; Pen of 5 Ewes and Pen of 5 Shearling ewes) contained 30 exhibits, and those of " Southdown or other Short-wooled Sheep " had 34. The remaining class for Long-wooled Sheep (presumably other than Leicesters) produced 17 entries, mostly the local Cotswolds or sheep of similar type.

The pigs were not classified according to breed. The show consisted of eight boars, four sows and four pens each of three young animals.

Names of many famous breeders appear in the award sheet. Apart from Bates, who won four of the five prizes for Shorthorns (and thus netted £70 by his adventure) we find Jeffries and Hewer winning with Herefords ; Bates's performance with Shorthorns was almost equalled by that of Mr. C. Large with his Cotswold Sheep. Mr. Shaw Lefevre won the Boar class. The total entries in the Live Stock Section numbered 213, and the whole exhibit was considered by many visitors to have been very remarkable. Mr. Drewry, who was one of the Tavistock party, says, " The show was considered to be a wonderful one ; there had been nothing like it before, and many said there would never be another like it," but the writer in the *Quarterly Journal of Agriculture*, evidently a Scotsman who had seen a good many " Highlands " between 1822 and 1839, was not so greatly impressed. He had admired the Kirklevington Shorthorns, Mr. Jeffries' Hereford bull and some Devon heifers ; but thought that otherwise the quality of the cattle " was in no way remarkable." The sheep " were in general good, and proved a pretty extensive show," but the pigs were " neither numerous nor remarkable." The Society's own *Journal* says that " it must be admitted that, if a foreigner had come to Oxford expecting to see the best show of breeding stock which England could produce, he would have been led to form a very inadequate idea of the merits of the different sorts of live-stock bred in this country."

As an after-thought, the Committee had decided to offer two prizes of £50 each for the best fourteen bushels of white and of red wheat of the harvest of 1838, grown by the exhibitor. It was intended that the judges should select certain of the parcels for sowing during the following year, and that the awards should take account of the yield and quality of the produce. The offer attracted 22 entries and the Judges duly made their preliminary selection ; but the samples became so mixed through the handling of the crowds of visitors that the final test could not be carried out.

A single page sufficed for the list of exhibits in the Implement Section. Nineteen makers were represented, but many of these showed single implements only—a turnip-cutter, a draining-plough or a corn-dresser. The only large exhibit was that of Messrs. Ransomes, who " sent up their waggons laden with more than six tons of machinery and implements, the superior manufacture and variety of which commanded universal approbation." The firm was awarded the Society's Gold Medal for its display. It seems, indeed, that the implement catalogue must have been incomplete, for the *Journal* mentions as exhibitors several firms, such as Howard and Garrett, which do not appear in the list. Two items of interest were a dynamometer, for measuring the draught of implements, and a " scorcher." This last was a kind of flame-thrower designed to destroy weeds, and said to be capable of dealing with an acre a day with a fuel consumption of only a quarter of a ton of coal.

In modern times the annual meeting of members in the Showyard has come to be regarded as a minor episode in a busy week. This may be unfortunate. In any case, it is clear that the early Country Meetings were primarily intended to be meetings in the ordinary sense, while the Show of live stock and implements was something incidental and secondary.

The climax of the Oxford meeting came at four o'clock on the Wednesday afternoon, when 2,450 members and guests sat down to dinner in Queen's College quadrangle, which had been roofed over for the occasion at a cost of £800. The toast list was, according to the habit of the time, extremely long, but enthusiasm seems to have remained at a high level until the end. The guest of honour was Daniel Webster, the famous American statesman and orator, who paid a tribute to the high level of English farming and the value of the English example to the rest of the world. His eloquence made a great impression on all who heard him. Mr. Chalmers Morton, writing forty-six years later in the *Journal*,[1] records his very vivid recollection of the scene : ". . . the homely, kindly presence of the late Earl Spencer, our first President ; the sonorous voice of the Duke of Richmond who succeeded him ; Mr. Pusey's pale and anxious, rather absent-looking face ; Mr. Handley's hearty jollity ; Baron Bunsen's

[1] 2nd Series, Vol. XXI (1885), p. 612.

staid and placid countenance . . . a distinguished row, seated on the dais : Daniel Webster also, evidently a great power both mentally and bodily."

The proceedings terminated on the following morning with a sale of some of the prize-winning stock. Mr. Paull had £140 for his prize-winning Devon heifer, Mr. Crisp's Southdown ram made £30, and several other animals sold for prices that were noteworthy for the time.

The *Quarterly Journal of Agriculture*, remarking generally on the results of the meeting, says :

" As a first effort, the getting up of the Meeting was highly creditable to the English Agricultural Society . . . if they are capable of such exercise in infancy, what will they not be able to accomplish in future years ? "

Before the Committee of Management dispersed to their homes they were invited to make Cambridge the scene of the meeting in the following year.

Encouraged by the progress which had been achieved, the Committee resolved to petition the Queen for a Charter of Incorporation. On March 6, 1840, the President was notified that Her Majesty had been pleased both to grant a Charter to the Society under the title of the " Royal Agricultural Society of England " and to extend her Royal Patronage to it. The Charter was sealed on March 26, 1840.

CHAPTER III

EARLY PROGRESS, 1840–45

DURING its first few years the Society made considerable progress with the tasks that it had set itself, and its membership rapidly increased. The total had passed two thousand before the end of 1839, and this figure was doubled in the ensuing twelve months ; by 1842 it had risen to 6,500, and by December 1844 stood at 6,827. Prices of farm produce were now good, farming was flourishing and there was great enthusiasm for agricultural improvements.

Three days after the Oxford Show a meeting of Cambridge citizens and Cambridgeshire farmers and landowners was held in the Red Lion Hotel, Cambridge, after the market " ordinary " ; the proposal was made that the Society be invited to the town for its second Country Meeting. Jonas Webb, Mr. Adeane's tenant at Babraham, who was then nearing the height of his fame as a Southdown breeder, and who had just returned from Oxford, seems to have been the chief speaker in favour of the proposal. Not unexpectedly, as it seems, objection was raised by a leading Cambridgeshire farmer, Mr. Thurnall, who had been the chief protester at the foundation meeting in the Freemasons' Tavern, and who was now President of the rival Farmers' Central Society. He now returned to the attack ; he said he could promise the support of his friends only if the Royal Agricultural Society would agree to rescind its resolution prohibiting the discussion of political questions ; but it seems that he got little support, and the meeting proceeded with the election of a Local Committee to make the arrangements in consultation with the Society. In the interval Mr. Shaw seems to have found the Secretaryship too heavy a task for his spare hours, and a full-time secretary was appointed in the person of Mr. James Hudson. The second show was duly held, on July 15, 1840, and under the Presidency of the Duke of Richmond, on Parker's Piece, Cambridge.

We may avoid a good deal of repetition if we consider the Cambridge Show and the five subsequent ones together. The list of the latter, with the President in each of the years, is as follows :

1841	Liverpool	Philip Pusey
1842	Bristol	Henry Handley
1843	Derby	Earl of Hardwicke
1844	Southampton	Earl Spencer
1845	Shrewsbury	Duke of Richmond

Taking the live-stock section first, only minor alterations were made in the classification during this period. Special classes continued to be provided for the three breeds of cattle (Shorthorn, Hereford and Devon) which had originally been selected as worthy of such recognition. The Shorthorn classes were well filled at all the shows, while those for the other breeds varied in size according to the distance between their respective breeding-grounds and the place of the particular meeting. At Cambridge a prize was offered for the best yearling bull of each breed, making six classes for each. At Bristol no premiums were offered for bull calves, but second prizes were provided for aged bulls, while at Shrewsbury the bull-calf class was reintroduced. All breeds were treated alike, and the prize sheet for Shrewsbury read, in each case, as follows :

Class
1 To the owner of the best Bull, exceeding 2½ years   £30
   To the owner of the second-best   do.   £15
2 To the owner of the best Bull exceeding 1 year but not exceeding
   2½ years   £20
3 To the owner of the best Cow in milk   £15
4 To the owner of the best in-calf Heifer, not exceeding 3 years   £15
5 To the owner of the best yearling Heifer   £10
6 To the owner of the best Bull Calf, not exceeding 12 months   £10

The names of many men who made history in their respective spheres appear in the prize lists of these early shows : Bates of Kirklevington, Booths of Warlaby and Killerby, Jaques of Richmond and Forrest of Stretton with Shorthorns ; Jeffries, Hewer and Meire with Herefords ; and, with Devons, Quartly and Turner.

The one exception to the ordinary provision for breeds was at Southampton, where classes were included for Channel Islands cattle.

In each of these years a further series of classes was provided for " any other breed or cross," and we may get some idea, from the awards, of the degree of improvement that had been achieved with the other varieties of cattle. At Cambridge all but one of the prizes went to the Sussex, the exception being the winning yearling bull, which was a Suffolk—presumably of the old polled dun breed which was to be used in the foundation of the modern Red Poll. At Liverpool the Sussex again led, with five of the nine awards, the other winners being a Longhorn, two cross-breds and a "Wyersdale" cow from Westmorland. At Bristol the Sussex had only one win, while the Longhorn had three and the local Old Glamorgan breed two ; the remaining two prize-winners are described as " Mixed or Cross-bred." At Derby, which was in the Longhorn's home country, this breed took all but one of the awards, the other going to a Shorthorn-Ayrshire cross. At Southampton, Sussex, Longhorns and cross-breds shared the prizes, while at Shrewsbury the Longhorns won everything, four of the five prizes going to the herd of the Hon. M. W. B. Nugent, of Higham Grange, Leicestershire.

Turning to the Cart-Horse section, classes were provided at Cambridge only for stallions and for mares with foals. At Liverpool second prizes were added, and at Bristol there were additional classes for two-year-old colts and for two-year-old fillies. One further class, for three-year-old stallions, was added at Shrewsbury. No breeds are mentioned in any of the award lists, but it is clear from the names and addresses of the exhibitors that most of the animals shown belonged to the one or other of the strains that were later amalgamated in the Shire Stud Book. On several occasions, however, Suffolks, from such well-known early studs as those of Crisp and Catlin, beat the Shires.

At Liverpool there were added, evidently as an afterthought and in connexion with a ploughing match that was held at the same time as the Show, classes for pairs and single horses " at plough." A similar event, arranged by the Local Committee, had taken place at Cambridge in the preceding year.

The only other type of horse provided for was the Thoroughbred, a premium of £30 being offered at each Show, except Liverpool, for the best stallion " which shall have served mares (during the preceding season) at a price not exceeding three guineas." The premium had been withheld at Oxford on account of the general lack of merit, but in each of the other years the award was made.

At each of the first five shows classes were provided for three groups of sheep, viz. :

(1) Southdown or other Shortwool.

(2) Leicester.

and (3) Other Longwool breed.

Among the winners in the Shortwool classes no breed is specifically mentioned except the Southdown, but it is known that some Shropshires appeared at Shrewsbury. Jonas Webb was easily the most successful exhibitor. At Liverpool, for example, he had three of the five prizes and his son took another, while at Derby the flock won four out of the six. The Duke of Richmond's Goodwood flock took the next largest number of awards.

The prizes for Leicesters were more widely distributed, but the most successful flocks were those of Pawlett of Beeston, Biggleswade and his neighbours on the light land of Bedfordshire. This district, which is now given over so largely to vegetable growing, was still, in the early years of the ' Royal,' " sheep-and-barley " country.

The prize-winners in the third group of classes, with the exception of a solitary Leicester-Lincoln cross, all came from a small area about Burford, North Leach and Charlbury on the borders of Oxfordshire and Gloucestershire. The sheep are variously described as New Oxfordshire, Oxfordshire Longwools, or Cotswolds, and it was by the last name that the breed finally came to be known. The most successful flock was that of Charles Large at Broadwell, though other well-known names—Handy, Edward Smith and Wells—appear in the lists. It is a rather melancholy fact that only a single flock of this breed, which at one time was in world-wide demand, now survives in its home district.

At Southampton in 1844 additional classes were provided for Short-wools other than Southdowns. The judges could find no shearling ram worthy of the prize offered, but awards were made to a Hampshire two-shear ram and to a pen described as " Berkshire Shortwool " ewes, which were probably of what came to be known as the Hampshire type. At Shrewsbury, in 1846, classes were provided for " Sheep of Mountain Breed " ; all the winners were Cheviots, and three of the four came from the flock of Robson of Kielder, near Hexham.

The grouping of the sheep, according to age and sex, varied little during this period. There was regular provision for shearling rams, for older rams and for pens of five shearling ewes. In some years a further class was made for pens of five ewes with lambs. The prizes were substantial, £30 being the regular figure for the first prize in the ram classes.

At Cambridge and Liverpool no attempt was made to segregate the breeds of pigs, prizes being offered for boars, for sows, and for pens of three gilts (from a single litter) of any breed or cross. At Derby the section was split in two, with parallel classes for the large breeds and the small. The breeds of pigs, in those days, were of course very ill defined, and the exhibitors rarely hazarded any description of their exhibits. From the rare cases where information is given, we gather that the Lincoln-shire, Leicester and Berkshire breeds were regarded as large, that there was a small Yorkshire and that the Improved Essex, the Dorset and the "Wadley" were also among the lesser sorts.

During the early years the implement section of the Show increased by leaps and bounds, and the quality of the exhibits rapidly improved. With a few notable exceptions, the exhibits at Oxford were "crude, cumbrous and ill-executed machines, the work of village ploughwrights and hedge-side carpenters." In a few years there was a great change. By 1841 Josiah Parkes could point to the "vast stride that had already been made in the mechanics of agriculture."

At Oxford there had been nothing in the nature of a competition, but a gold medal was given to the firm of Ransomes for its collection, and silver medals were awarded to three other exhibitors for a ridging plough, a drill and a corn-dressing machine respectively. At Cambridge a special premium of £20 was offered for a gorse-crushing machine intended to make gorse a possible winter food for stock—a strange selection from among the various desiderata in the way of farm machinery. The Cambridge prize-sheet also gave out that such sum as the Society might think proper would be awarded to any new agricultural implement of merit. Nobody produced a gorse-crusher that satisfied the judges, but seven medals were awarded for new implements, among which we may note a tile-making machine (turning out the old pattern of drain tiles with a horseshoe-shaped upper portion and a separate sole), two drills, a stubble rake, a liquid-manure cart and Crosskill's clod-crushing roller. Various implements were put to work, but there was nothing in the nature of tests; the judges merely inspected the exhibits and bestowed their awards as they thought fit.

At Liverpool the prize for a gorse-crusher was offered again, and was awarded. Moreover, there was an elaborate prize-list, and also some attempt at a trial of the exhibits in several of the classes. The judges awarded forty money prizes, ranging from £2 to £25, for ploughs, cultivators, drills, horse-hoes, chaff-cutters, cake-crushers, corn-dressing machines and miscellaneous tools. The trials were carried out on the race-course at Aintree and in the presence of a large number of members and others, and were completed in a single day. The only thing of the nature of a test or measurement was one of the draught of the ploughs, carried out with a rather crude sort of dynamometer. Even this, however, was enough to show the wide range of variation; a Ransome and a Howard plough, along with two others, showed a draft of 28 stone, while others ran up to 40. Most of the leading implement makers seem to have competed; besides Ransomes and Howard there were Wilkie of Uddingston,

Farmers' Journal.

THE ACCREDITED ORGAN OF THE NOBILITY FOR THE PROTECTION OF AGRICULTURE. THE DUKE OF BUCKINGHAM AND CHANDOS PRESIDENT

INTERIOR OF THE CATTLE YARD AT PARKER'S PIECE

DINNER OF THE COMMITTEE IN THE HALL OF TRINITY COLLEGE

GRAND DINNER IN THE PAVILION AT DOWNING

PLOUGHING MATCH SKETCHED FROM THE GREAT SHELFORD ROAD

IV.—SCENES AT THE CAMBRIDGE MEETING, 1840

VA.—Francis Quartly with his Favourite Devon Cow
From a presentation portrait of 1850

VB.—Jonas Webb and his Southdowns, 1841
From a painting in the possession of the Society

near Glasgow ; Garrett of Leiston ; Hornsby of Grantham ; Smith of Deanston (the great advocate of covered drains and subsoiling) and the Earl of Ducie, exhibiting on behalf of Morton of Stroud. The greatest innovation, a portable steam-engine shown by Ransomes, received only a commendation.

At each of the four subsequent shows the system was the same : a prize sheet was published many months before the Show, money prizes being offered for a long list of specified machines as well as for ' any other ' implement ; in each year there was some attempt at a trial ; and in each case a long and detailed report was drawn up by the judges or, later, by Josiah Parkes who, in 1843, was appointed consulting engineer.

Year by year, as the number of entries rose, the task of the judges became more difficult. At Derby in 1843 they felt obliged to withhold the awards in certain classes because they felt that " more lengthened and accurate trial " was necessary if a just decision was to be reached : they further demanded that a proper catalogue of the entries should be drawn up, and that the public should be excluded from the trial ground until the judges had finished their work. Accordingly, at Southampton in 1844, the public was excluded, and a more thorough series of tests was carried out by three separate groups of judges ; these tests disclosed a great many defects, and the judges were severe in their criticisms. In 1845 the judges made further suggestions, particularly for deferred tests of those implements which could not well be put to work upon a soil that might be hard baked under a July sun.

But such minor improvements in the system of awarding premiums were not enough. The whole affair was becoming unwieldy ; and the trials, useful as they were in the first few years in distinguishing the apparently good from the obviously bad, were not calculated to distinguish between the finer degrees of merit. A new scheme was felt to be necessary, and this was devised, a little later, through the joint efforts of H. S. Thompson (Sir Harry Meysey Thompson) and C. E. Amos, the latter of whom, in 1848, replaced Parkes as Consulting Engineer. The further progress of this side of the Society's activities is described in Chapter VII. Meantime, we may glance briefly at some of the major improvements which the judges recorded in the years with which we are here concerned.

The portable steam-engine shown at Liverpool was the first of many. Two years later the judges could say that " the manufacture and use of travelling steam-engines is become a systematised business " and those exhibited at this and subsequent shows were regarded as very fairly efficient ; such engines, it should be noted, were not locomotives—i.e.

they were merely mounted on travelling wheels and were hauled from one place to another by teams of horses. Drills improved very quickly, the makers of the less satisfactory types copying the mechanisms of those that were seen to be better. The first " haymaker " (tedder) appeared in 1843 and was considered a practical tool. An efficient dibbling machine had been exhibited, and in 1845 a turnwrest plough of excellent design was awarded a prize.

At this time the cost of drain-tiles, made as they were by hand moulds, was one of the great obstacles to large-scale land drainage, and there were continuous efforts to evolve tile-making machinery. An award was made to Ransomes in 1844, but the greatest step of progress was the machine of Thomas Scragg of Tarporley, Cheshire, which was awarded the twenty-pound prize in 1845. Perhaps the most exciting of the various new inventions which failed to gain prizes was Lady Vavasour's rotary tiller, of her own invention, shown at Bristol in 1842. This consisted of a large barrel into which were fixed a large number of spoon-shaped teeth ; it was, like other tillers and diggers of later years, to act as a substitute for plough, harrow and cultivator. Unfortunately, upon trial, " the earth adhered to the teeth, closed the spaces between them and accumulated until the machine became an immensely heavy roller, consolidating the ground and completely reversing the intentions of her ladyship."

There is no doubt that the implement exhibits at the early Shows resulted in great benefits to the industry. It may have been that the judges could do little to instruct the better makers, but the makers learnt much one from another ; and although the best makes of ploughs or drills in 1845 may have differed but little from those of 1839, the general standard of design and workmanship improved immensely.

The Society had from the outset wished to do something for the improvement of crops, and they naturally turned their attention in the first instance to wheat, which was by far the most important crop in the England of a hundred years ago.[1] As already recorded, the competition at Oxford had miscarried ; the prizes were offered again at Cambridge, but it was provided that the bulk of each parcel was to be sealed and reserved, the public being left to make free with half a sack of each sort. The exhibits of red varieties were rejected as impure, but two white sorts were selected for field trial. In each of the three trials, however, local varieties proved to be more productive than the show seed, and the prize was therefore withheld.

[1] McCulloch's estimate of the Area of Wheat in Britain, at this period, was 4·15 million acres, or considerably more than twice the area grown in 1938.

At Liverpool in the following year two lots of white wheat and two of red were selected, according to plan, and arrangements were made for three separate trials. Unfortunately 1842 was a bad wheat year ; Mr. Handley's plots in Lincolnshire were practically a failure, while Mr. Kimberley reported that his trial, owing to the unfavourable season, had been " unsatisfactory and inconclusive." The third of the trio of experimenters, Mr. W. Miles of King's Weston, near Bristol, had better luck, and wrote a full account of his experiment for the *Journal*.

The interest of this report lies in the excellent technique which was employed in the trial. Mr. Miles laid out the field in half-acre plots and carried out the experiment in triplicate. He included not only the four selected wheats from Liverpool but seven others, obtained from various sources, for purposes of comparison. He made careful observations of the plots at various stages of growth, and gives information about weather conditions. A sample of one tenth of an acre from each plot was harvested by scythe, threshed by flail and the produce, both grain and straw, was carefully weighed. Each lot of grain was separately ground, and baking tests were carried out with the flours. Colonel Le Couteur's beautiful white sort, a selection from Talavera, failed during the winter, and the yield from the second selection of the Liverpool judges, a lot of Chidham, was only moderate. The Burwell red wheat from Liverpool was the poorest of its group, and though the other selection of the judges gave six bushels an acre more than any of the others, the Council declined to award the prize.

Mr. Miles, naturally, did not make a mathematical estimate of the probable error of his results, but otherwise his method was above reproach, and it is a matter for regret that he did not pursue his career in field experimentation and turn his attention to some more promising problem. The wheat trials were continued for a year or two more and then dropped. It is clear that, as originally planned, they were not worth carrying on. Even with an elaborate system of testing, involving trial stations widely distributed over the country and care to ensure purity and constancy of each strain, it is difficult enough to distinguish between the better and the less good varieties of corn. The difficulty was not appreciated when the Society started its scheme.

Albert Pell, who later became a great figure in the agricultural world, wrote some entertaining reminiscences of these early Shows. He was still an undergraduate at Trinity when the Show came to Cambridge, and confesses that he was less concerned for its success than about the state in which it would leave the cricket-pitch on Parker's Piece. By 1842 he had graduated and, having developed an interest in farming, he decided to make the journey to Liverpool by the new railway. Pickpockets abounded in the Showyard, and Pell's companion, who carried their common purse, was among their victims. " I was left," he says, " with a few shillings, and in debt to the Adelphi Hotel, and returned at once by night train to London, wet through, in a carriage with no glass but with a substitute in the shape of louvreboard to improve the draught and increase the gloom. This journey served me for two years and I left Bristol alone, but recovered my spirits sufficiently in 1843 to venture to Derby." There he saw his first portable steam-engine, exhibited by Dean of Birmingham. . . . " I think it took a prize. I know it took in a cousin of mine who, as the daring purchaser of this tremendous novelty, became the hero of the Showyard for a few hours."

At Southampton in 1844 Pell made a purchase of a set of harrows from James Howard, a member of the already famous implement firm who was also destined to play a large part in the affairs of the Society. . . . " I have a remembrance of a fresh-coloured young man meeting me as I drew near these zigzag tormentors of the soil, and pointing out their charm with such insinuating effect that I became their purchaser on the spot. This is how we got the harrows from the Royal Showyard to Ely in Cambridgeshire—first by rail to Nine Elms (London), next to an inn in the Old Bailey, the terminus of a carrier's cart which got them to Cambridge, whence another of Hobson's calling transferred them to Ely. They arrived in time for the autumn wheat-seeding."

By this time, having fairly embarked on farming, Pell did not like to miss a Royal Agricultural Show, so betook himself in 1845 to Shrewsbury. There was still no line beyond Wolverhampton, the occasion being, in fact, the last on which a Royal Show was carried through without railway facilities. At Wolverhampton he contracted with a coach-driver to be taken to the Showground for ten shillings, and climbed to an outside seat ; a remarkably well-dressed gentleman, much inclined to conversation, sat on the one side and a very plain, silent personage on the other. " The left-hand man was for knowing nothing, not even what his destination might be. The right-hand one knew everything and something more— was intensely interested in farming, specially animals, but was curiously in doubt whether a ' lamb hog ' was a sheep or a pig. At last we pulled up at the Raven Inn, where stood two stoutish gentlemen to whom my silent friend nodded, at the same time saying with an air of authority to my other neighbour, ' You stay where you are.' There being a little commotion at the door of the omnibus, I looked round and beheld three more smart fellows alighting into the charge of my silent friend's familiars. The four swell-mobsmen were seen out of Shrewsbury forthwith and we were advised to examine our purses." Going to the ' Royal ' was thus still something of an adventure.

During this period there were three meetings annually of the Society's members. Two of these, in May and December respectively, were held in London. At each the members received a Report from the Council on its activities during the preceding half-year and considered the accounts for the period. At the December meetings it was usual to have one or two scientific lectures, Professor Daubeny of Oxford beginning the series and Dr. Lyon Playfair continuing, Playfair having been appointed in 1843 to be the Society's Honorary Consulting Chemist. At the country meeting there seems always to have been an " ordinary " on the first day, after which papers and essays were read, and a banquet on the second afternoon. The cost of accommodating the thousands of diners at Oxford and Cambridge had however proved excessive, and therefore some limitation was imposed at subsequent Shows.

The Council held a monthly business meeting in London on the first Wednesday of each month, and these meetings were not open to ordinary members ; but meetings seem to have taken place almost regularly on other Wednesdays, were open to all members and took the form of conferences and discussions on general agricultural topics.

The Council had a good deal of work, in its early years, in revising and extending the Society's by-laws, and in drawing up regulations for the exhibit of live stock and implements. In 1841 it decided upon a regular itinerary for the Show, dividing England and Wales into nine divisions which were to be visited in turn. The object, of course, was to ensure that each part of the country should have a fair share of the Society's attention.

In 1841, also, the Council looked round for suitable permanent quarters and asked the Crown Lands Department to provide a site on which a permanent building, specially designed for the purpose, could be erected. A site in the neighbourhood of St. Martin's Lane was offered and plans were got out ; but after consideration the Council decided that the cost would be more than the Society could afford, and the scheme was abandoned. Instead, the Council secured a ninety-nine years' lease of No. 12 Hanover Square at a rent of £330. A considerable sum was spent on the adaptation and improvement of the house ; the basement was fitted up as a museum and store-room of implements, models, varieties of grain, etc., and, besides offices and a Council Room, there was a Library, a Reading Room and living accommodation for the Secretary.

In 1844 the Society had the opportunity of acquiring the library of the old Board of Agriculture. More than twenty years earlier, at the dissolution of the Board, the library had been bestowed upon the father of Mr. Webb Hall, whom we find taking a prominent part in the arrangements for the Bristol and Southampton meetings. Mr. Webb Hall died in 1844, and his executors agreed to hand over, at valuation, the collection of books to which so great an historical interest attaches.

The early years of the 'Royal' saw the introduction of three new fertilizers—nitrate of soda, superphosphate and guano. The Council encouraged members to carry out and record experiments with these, and a large number of members' reports appear in the early volumes of the *Journal*. The first number of the *Journal* appeared in May 1839, and the intention was to publish three or four times a year ; but the cost of distribution proved excessive, and from 1843 onwards it was issued in half-yearly parts. The responsibility for the publication was placed in the hands of a Committee, with the Secretary as Editor ; but from the outset the real direction was in the hands of Philip Pusey. The first article, "Some Introductory Remarks on the present state of Agriculture as a Science," is by Pusey himself. The material for the *Journal* in its early days consisted largely of papers read at the Society's meetings, and of prize essays. The number of prizes offered for such essays was generally ten in each year, and the money awarded amounted to some £300 per annum. The story of the *Journal* is told in Chapter XII.

Turning to the financial history of these years, the Society seems to have expected a regular deficit on its annual Shows ; at any rate there was a regular excess of show expenditure over show income, and the Council expressed no anxiety on this score until 1843, when they confessed that they were somewhat alarmed by a loss of £1,700 at Derby. The following figures tell how the deficit tended to rise as the scale of the Show enlarged. Southampton began the practise of a local contribution to the expenses of the Show, and the Council expressed its appreciation of the town's very generous gift of £1,000. Shrewsbury followed this excellent example by providing an equal sum, but the lack of a railway meant a small attendance, and the deficit rose again (see table on page 37).

Despite these Show deficits, however, the Society's total income exceeded its expenditure in each of the first six years, and it raised its reserves (apart from funded life subscriptions) to a figure of the order of £5,000. But the margin of income over expenditure became a slender one in 1844, and there was a deficit of over £400 in 1845. It was obviously

	Stock Entries	Implement Entries	Expenditure	Receipts		Excess of Expenditure
				Direct	Local Fund	
			£	£	£	£
1839 Oxford . . .	249	23	2,688	2,394	—	294
1840 Cambridge . .	352	36	3,589	3,416	—	173
1841 Liverpool . .	319	312	5,052	4,106	—	946
1842 Bristol . . .	510	455	4,775	4,202	—	573
1843 Derby . . .	730	508	5,090	3,390	—	1,700
1844 Southampton .	575	948	5,736	3,929	1,000	807
1845 Shrewsbury. .	437	942	5,166	2,662	1,000	1,504

becoming necessary to consider whether to aim, for the future, at a self-supporting Show, or whether to restrict the expenditure on other objects.

In 1845 the Society had to lament the death of its founder, Earl Spencer, and showed its sense of his services by electing his brother, and successor in the title, to the vacancy in the list of Trustees.

CHAPTER IV

LIVE STOCK AND SHOWS, 1846-79

IN accordance with the scheme that had been laid down in 1841, the Society's country meetings for the seven-year period 1846-52 were held successively at Newcastle-on-Tyne, Northampton, York, Norwich, Exeter, Windsor and Lewes. The Windsor meeting of 1851 was something exceptional, on account of its connexion with the first International Exhibition ; the machinery section of the Society's Show was cancelled, in order that there might be no competition between the two events, and the live-stock Show was made larger than usual. Apart from this there were no important changes in the character of the Show or in the prize list. The number of animals exhibited, however, steadily increased and, by 1852, had reached about three times that which had appeared at the first show at Oxford.

Among cattle, the largest representation was always that of Short-horns ; the average number, over the seven years, exceeded a hundred head, and there was but little variation from year to year. The leading prize-winners were Richard Booth of Killerby, John Booth of Warlaby and Colonel Towneley, of Towneley Park, Burnley ; but the award lists show that pedigree herds were now becoming widely distributed. In 1851, for instance, Shropshire, Lincolnshire, Norfolk and Scottish herds were competing with those in the breed's home district in Durham and Yorkshire.

By contrast, the entries of Herefords remained rather small, and the prize-winning herds were still, for the most part, confined to the Welsh Border counties. The exceptions were those of Fisher Hobbs, near Col-chester, and of Lord Radnor, at Faringdon in Berkshire. The most familiar names in the Hereford world were Carpenter, Edward Williams, Aston, Price, William Hewer and Lord Berwick.

Entries of Devons were up or down according to the distance of the Show from the breed's home county ; only fifteen head went to Newcastle in 1846, whereas at Exeter, in 1850, there was a collection of over 120, including all the finest specimens of the breed. There was already strong rivalry between the breeders in Devonshire itself—Quartly, Davy and Turner—and the colony in the Bridgwater district of Somerset, which included Farthing, Fouracre and Hole. The most notable herd outside the South-West was that of the Earl of Leicester at Holkham, which had been founded by his father (Coke of Holkham) in 1791, and from which the first exports of Devons had been sent to the United States.

From 1846 onwards cross-breds were excluded from the fourth group of classes, these being henceforward confined to cattle " of any other pure breed." Here the Longhorn continued to take a majority of the prizes, though an occasional Galloway, West Highland and Ayrshire appears in the list and, at Norwich, a Norfolk Polled and a Suffolk Polled were respectively first and second in the aged bull class.

Apart from these regular classes, and leaving aside the Windsor Show of 1851, there was provision for South Hams (South Devons) at Exeter in 1850 and for Sussex at the Lewes Meeting of 1852. The South Devons could muster only ten animals, but the display of seventy-five head of Sussex at Lewes was very impressive—so much so that more than one visitor suggested that the breed now deserved a permanent section in the prize list. It was long, however, before such regular classes were provided.

At Windsor in 1851 a special effort was made to get together a completely representative collection of all the improved breeds. Separate classes were made for Longhorns, the Channel Islands breeds, Sussex, Scotch Horned, Scotch Polled and finally for " Welsh, Irish or other Pure Breed." The result was an exhibit which was at least good in parts. Thirty Channel Islands cattle appeared, variously described as Guernseys, Alderneys, or merely as " Channel Island " ; there was a good show of Sussex, though not so good as that of the following year at Lewes. Ayrshires won all the prizes offered for Scotch Horned Cattle and, in the Polled classes, William McCombie of Tillyfour won three of the four prizes with his Aberdeen Angus. This was the first occasion on which improved specimens of the Angus breed had been seen at an English show, and they seem to have made a favourable impression. In the remaining class the Welsh and Irish made no showing ; all the awards went to Red Polls, still shown under the old names of Suffolk or Norfolk Polled.

Turning to the sheep classes, the Southdowns and the Leicesters continued to be the most popular breeds, with the Leicester now generally producing the larger numbers. In the earlier of the seven shows the leading prize-winners were Pawlett of Beeston (Bedfordshire) and William Sanday of Holme Pierrepont (Notts.). Later on, however, we find a good many prizes going to the Yorkshire Wolds, which district was later to become the stronghold of the breed. Fisher Hobbs bred Leicesters as well as Herefords, and Turner of Barton, in Devon, had a Leicester flock besides his noted herd of Devon cattle.

Jonas Webb continued to win the prizes for Southdown rams with a monotony that his rivals must have thought depressing ; but it was against his principles to make up his ewes into the very high condition that was then necessary to the winning of prizes, and he left the female classes to others. Of these the most successful was the Duke of Richmond, who had a very large and choice flock at Goodwood. The name of Fisher Hobbs, that most versatile of all the great farmers of the time, also appears occasionally in the lists.

" Other Longwool Breed " still meant, for practical purposes, Cotswolds, and the familiar names—Large, Garne, Hewer, Lane, Handy and Edward Smith—kept reappearing time and again.

At Newcastle the Society provided classes for sheep " of mountain breed " and all the prizes went to Cheviots. At Lewes, in 1852, a special section was made for Romney Marsh, and this drew a good entry, though " the breed was not highly spoken of by the judges."

The original classification of horses—according to type rather than breed—was retained during the period. " Agricultural Horses " included Suffolks as well as the various local strains which were later brought together in the Shire breed. The Suffolks had most of the honours. The " Agricultural " section was the only one that was maintained regularly throughout the period, though, from 1848 onwards, there was always a class for hunter stallions and another for roadster sires. At York in 1848 there was a good display of Cleveland Bays, which were then still largely bred for stage-coach purposes.

Pigs were still grouped as large or small, though there was an increasing tendency on the part of exhibitors to specify the breed to which their animals belonged. Among those mentioned are the Leicestershire, the Improved Essex, the " Essex-enlarged," the Berkshire and the Tamworth. Fisher Hobbs, with his Essex, won perhaps more prizes than any other exhibitor, but the most interesting development of the time was the entry into the field of a group of Yorkshire weavers, pre-eminent among whom was Joseph Tuley, of Exleyhead, Keighley. Tuley made his first appearance at York in 1848, when he won two first prizes ; encouraged by this success, he took two sows all the way to Norwich in 1849, and another to Exeter in 1850, and brought home first prizes for all three. He again had first prizes at Windsor for a sow and for a pen of three gilts, and it was now recognised that a man who was ekeing out his eighteen shillings a week by keeping a few pigs in his back yard was in fact a master stockman. In the 'fifties and 'sixties pig breeding became the absorbing interest of great numbers of weavers round Tuley's home, and the pig judging at Keighley Show drew great crowds. To Tuley belongs a large share of the credit for the creation of the modern Large White breed, which was later to provide the most important raw material for the world's bacon industry.

In 1852, at Lewes, a show of poultry was held for the first time. This was continued for six years more, but did not become a permanent institution until much later.

The period from 1853 till 1861 was one of increasing prosperity in British agriculture, but interest was concentrated on arable farming and corn-growing rather than on live stock. Tile-draining and steam-engines were, for the moment, of more importance to the general run of farmers than cattle and sheep. So far as concerns stockbreeding, it was a time of quiet and steady progress rather than of spectacular happenings. The Kirklevington herd of Shorthorns had been dispersed in 1850 (following Bates's death) and the average price realised was less than £70. Earl Ducie's sale of very select Bates Shorthorns followed three years later, and a few American buyers, with long purses, were in attendance. A "Duchess" cow went to the United States at seven hundred guineas, and two bulls left the country for the same destination at 650 and 500 guineas respectively. The average for the whole collection was just over £150, or practically the same as that realised at the great Ketton sale of 1810. The sale may be said to mark the beginning of the regular trade in high-class pedigree cattle to the new countries.

Up till this time most of the highly improved live stock still traced back to the herds and flocks of a very few early improvers, and the numbers of really choice herds and flocks were very small. The Booth herds had been derived direct from those of the Brothers Colling, and the Booth blood was now generally recognised as representing the best of the breed. Cattle of this strain were winning three-fourths of the prizes at the leading shows, and were being used as the means of cattle improvement in most of the beef-producing areas, in Scotland and Ireland as well as in England.

The Hereford, which owed much to the early work of Tomkins and Price, was now a well-fixed breed, with the colour pattern which we now know ; its Herd-Book had been started in 1846 and, by the 'fifties, there was already an export trade to America and Australia. Devons had been among the earliest exports to the United States, and pedigree registers were opened, both in the old and the new homes of the breed, soon after the middle of the century.

The best draft horses of the Shire type were bred in South Lincolnshire and Leicestershire, and the finest Suffolks in the Woodbridge district.

Both the Babraham and the Goodwood Southdowns were direct descendants of the original improved flock of John Ellman at Glynde, though selection had led to some increase in size. The owners of both were already selling for export to France, Prussia and Russia. In 1855 Jonas Webb showed sheep at the Paris exhibition, and the French Emperor so admired one of the rams that Webb felt obliged to make him a present of it ; it was an expensive gift, but it proved an excellent advertisement for the breed and for the Babraham flock.

The Beeston and Holme Pierrepont flocks of Leicesters, both of the purest Dishley blood, continued during the 'fifties to take between them five out of every six prizes that were offered for the breed ; their main market was still at home, but exports had begun.

The years round 1860 were the hey-day of the Cotswold breed ; at that time as many as five thousand rams might be let at Gloucester Ram Fair alone. The breed acquired fame both in Australia and in America, and many long prices were obtained for export. In 1861 Lane of North Leach sold the group of rams shown in the accompanying picture at prices ranging from 80 to 120 guineas each.

A good many other breeds, both of cattle and sheep, were now beginning to take definite shape, and were in course of improvement, so that the Society in the future was often to be faced with the problem of deciding whether one or other of these was sufficiently well fixed in type, was good enough, and was important enough in regard to numbers, to be worthy of recognition in the form of special classes at the Show. In the 'fifties the future of many of the newer local breeds was still obscure, and the general policy of the Council was to offer prizes for any breed which was extensively kept in the district where the meeting of the year happened to be held, and which could put up some kind of claim to be both ' distinct ' and ' improved.'

Under this plan there were classes for Welsh Cattle at Gloucester in 1853, and £70 was offered in prizes. But in this case the move seems to have been premature, for only five animals were shown and these, in the opinion of the cattle steward, " were not worth the amount of the prizes " ; he indicated, indeed, his clear opinion that only three breeds of cattle were truly worthy of a place in the Society's Shows. On the other hand, the new class for Shropshire Downs, first provided in the same

VIA.—ROBERT LANE AND HIS COTSWOLD RAMS
From a painting by G. R. Whitford, in the possession of Mr. Wm. Garne

VIB.—LEICESTER ONE-SHEAR WETHER
From Low's " Domestic Animals of the British Islands," 1842

VIIA.—BERKSHIRE PIG
From Low's " Domestic Animals of the British Islands," 1842

VIIB.—THE BRISTOL MEETING, 1842
From a print in the possession of the Society

year, was very successful, and the stewards expressed the hope that " the Society would recognise them as a distinct breed."

The most interesting feature of the very successful meeting at Lincoln in 1854 was the collection of ' Improved Lincoln ' sheep. Lincoln breeders had long had the opportunity of competing in the class for ' Longwools other than Leicesters ' but had left the field to the Cotswold men. Now, when they were offered special classes, they produced some excellent sheep. The steward's report, however, indicates that the breed had still some leeway to make up before it would be able to compare favourably with the Leicester and Cotswold.

At Carlisle in the following year there were classes for Ayrshire, Scotch Polled, and Scotch Horned cattle. " In these classes the quality of the few animals exhibited was very bad "—excepting only four excellent Aberdeen Angus exhibited by McCombie. Mr. Simpson, the reporter, rather gave up his attempts at description when he came to the mountain sheep classes—" The judges seemed to consider some of the animals superior, but not understanding their merits I can only remark upon them as being, in some instances, extraordinary-looking creatures with small bodies and enormous horns." As regards pigs, " The numbers were far below average and a great many, I regret to say, much above the age stated in their certificates."

In 1855 was held the Paris exhibition, and a large exhibit of British live stock was sent over. In the following year the French Government reciprocated by asking the Society to provide classes for French cattle at the Chelmsford Show of the year. The Council voted £150 as prize money, and the result was a good exhibit of the Charolais, Norman and Brittany breeds, which naturally created much interest.

The first permanent addition to the list of live-stock classes was made at Salisbury in 1857, when a fourth section was added to the sheep. This was for " Shortwools other than Southdowns " and the contest resolved itself into one between the Shropshire and the Hampshire. Honours were even, with three prizes each. Among the winners we find the greatest of the early breeders—Humfrey of Oakash, Wantage, with Hampshires, and Samuel Meire of Castle Hill, Much Wenlock, with Shropshires. At Chester in 1858, where the collection of live stock was exceptionally large and good, there was a flood of competition in the new class. There were sixty-one entries in the aged-ram class, and these included Oxfords, West Country Downs and Cheviots besides Shropshires and Hampshires. The Shropshire had now definitely emerged from its position as a purely local breed, demand was increasing, and at some sales rams had reached average prices of the order of £20 each. Robert Smith, the steward, also remarked on the noteworthy improvement in the Hampshire. He mentioned the growing importance of the Lincoln in its native county, and regretted that the Lincoln breeders were still refusing to compete with the Cotswold men.

In the following year Smith again urged the Lincoln men to display their wares, and added the Romney Marsh to the list of longwools that might be expected to compete.

At Warwick in 1859, the steward expressed his sympathy with the judge in having to place large classes of sheep containing the Hampshire or West Country, the Shropshire and the Oxford breeds. In the event, the Hampshires had the bulk of the honours, Humfrey himself carrying off four of the nine money prizes. From Canterbury onwards the immediate difficulty was met by providing special Shropshire classes, and for some years the Hampshire men were left with the " other Shortwool " class practically to themselves. The Shropshire breeders took full advantage of their opportunity. Mr. Randall, the judge at Leeds in 1861, wrote :

It is only within the last eight or ten years that Shropshire sheep have come prominently into notice. Yet here (at Leeds) we found them in greater numbers than any other breed. . . . It is impossible not to be struck with the appearance of these sheep as a most useful, rent-paying kind of animal ; and if they have not yet attained that uniformity of character which we are accustomed to see in some other breeds, it must be admitted that they constitute a near approach to perfection ; all that the breeders have to do is to concentrate these qualities by careful and judicious selection.

The class for " other recognised breeds " of cattle was now presenting a similar problem. At Chester in 1858 the entries comprised 2 Alderneys, 4 Ayrshires, 4 Polled Norfolks, 5 Polled Angus, 2 Welsh, 3 West Highlanders, 1 Galloway and 1 Brahmin (Indian Zebu). At Warwick the judge distributed his awards among two Angus, and one each of the Welsh, Jersey and Norfolk Polled. His aged-bull class presented him with a choice between one each of the Sussex, Welsh, Angus, Alderney and Norfolk breeds.

At Leeds in 1861 the Sussex was given classes of its own, but the position was to require continual review in future years.

During this period the Council had to turn its attention to a number of rather general questions in connexion with the exhibit of live stock. One difficult problem arose out of the growing practice of over-fattening the cattle, sheep and pigs. There were, and have continued to be, two sides to the question. On the one hand, it is obviously undesirable that a breeding animal should be brought, for any purpose, to a pitch of fatness that is likely to interfere with its capacity to breed ; on the other, the capacity to lay on fat is an essential characteristic of the meat type, and it is often impossible for a judge to distinguish between the animal that is lean because it has not been forced and another which is lean because it will not fatten.

In 1853 the Council tried the experiment of appointing special juries to make a preliminary examination of the stock, and the juries were instructed to disqualify any animal which, in their opinion, was in over-fed condition. That this responsibility was too great was proved by a sow which, after having been disqualified for gross overfeeding, proceeded to give birth to a litter of eight fine healthy offspring. After a trial the Council reverted to the earlier plan of leaving the decision to the good sense of the Society's judges.

In 1861 Mr. Fisher Hobbs, as chief live-stock steward, had some remarks to make on the new practice of providing foster-mothers for the show cattle. Sixteen nurse cows were accommodated at Leeds in that year, and one " calf " which he found among the exhibits had reached the rather mature age of seventeen months. He suggested that some definition of a calf was called for ; but this was another subject upon which it was very difficult to legislate.

In the early years of the Show there had been frequent cases of serious unsoundness among the horse exhibits, but the practice of having a regular veterinary examination by the Society's veterinary inspectors, and of disqualifying any that showed hereditary defects, had led to rapid improvement. In 1861 Professor Spooner reported not only that the general quality of the horses shown was rapidly improving but that cases of unsoundness were becoming rare.

As is mentioned later (Chapter VIII), another difficulty in the early years was that of checking the declared ages of the animals exhibited. One gathers that the standard of morality was lowest among the pig fanciers. But Professor Spooner soon became an expert in the detection of crime, and the disgrace of a public exposure, upon which the Council insisted, was a very effective deterrent.

The Society was enabled to extend the show prize sheet, during the late 'fifties and the 'sixties, by reason of the growing popularity of the Show with the general public. The Chester Meeting of 1858 was the most noteworthy financial success in the Society's history, the profit passing the thousand-pound mark. At Leeds in 1861 arrangements were made for parades of live stock, it being felt that the ordinary visitor had hitherto been given no sufficient opportunity of seeing the animals. The parades added a great deal to the popularity and educational value of the Show, and were retained as a permanent feature.

The second of the Society's special efforts was made in 1862 at Battersea, the year of the second great International Exhibition at Hyde Park. A shadow was cast over the preparations by the death, in December 1861, of the Prince Consort, who had been elected President for the year and whose interest in agriculture, and in the Society, was very real.

There was a wide extension of the prize list, the customary classes being in many cases subdivided, and provision being made for a large number of the less important breeds. Classes were provided for Longhorn, Welsh, Kerry and Channel Islands cattle ; for Suffolk horses ; for Lincoln, Cotswold, Hampshire and West Country Down, Dorset Horn, Oxford Down, Kent or Romney Marsh and Mountain sheep ; and finally for Berkshire pigs. This list, with that of the breeds formerly recognised, may be taken as a complete one of the improved breeds to be found in England and Wales at the time.

The Highland and Agricultural Society brought their annual show to the Battersea ground, and offered prizes for the Scottish breeds. Finally, in order to give the Show an international character, a sum of £750 was provided as prize money for foreign live stock. The total entries of live stock reached 1,986, or nearly double the ordinary number of the preceding years. Of these, 1,565 came from England and Wales, 238 from Scotland and 183 from foreign countries. An innovation was the judging of the live stock in public, and although the price of admission for the judging day was fixed at a guinea, in order to avoid the risk of an inconveniently large crowd, well over a thousand people paid to see. France, Switzerland and Holland sent cattle, and the sheep included a wide selection of various strains of Merinos, from France, Spain and Saxony. The features among the English classes were those for Shorthorns (of which there were 250 entries), Red Poll cattle and Suffolk horses.

The weather was fine, but the attendance, though it reached nearly 125,000, actually fell short, by 20,000, of the total at Leeds in the previous year. The Society was to have further happy experiences of the interest in agriculture of the northern industrial towns and a further (and almost fatal) experience of the lack of interest on the part of the London public. It was never expected that the Battersea Show would cover expenses, but the deficit of nearly £3,700 was the largest experienced up till that time, and represented more than a fifth of the Society's reserve funds.

Many of the new and minor breeds had produced, at Battersea, unquestionable evidence of their right to a place in the sun of the Society's favour, and the succeeding Shows were much more representative of the now rapidly spreading tendency towards the systematic improvement of the whole mass of the country's stock.

New men also began to appear among the exhibitors of the old-established breeds, and some of these were destined to play important parts in the history of British live stock. Richard Booth and Colonel Towneley still appear among the Shorthorn prize-winners of 1863 to 1865 ; but other breeders of the Booth type were strongly competing ; moreover, Lord Feversham and others were returning to the attack with their Bates strains ; Richard Stratton was winning prizes with his South-country dual-purpose type and, in 1864 at Newcastle, an obscure farmer called Amos Cruickshank, of Sittyton in Aberdeenshire, won the old-bull class.

With Leicesters, Sanday and Pawlett were still to the fore, but John Barton, of Malton in Yorkshire, was now the leading prize-winner, and numbers of other pedigree flocks in the Yorkshire Wolds were making progress.

Jonas Webb had sold his famous Southdowns in 1861, the flock of 1,404 head realising £16,646. He died the following year—probably the most prominent figure in English farming, and widely known in many foreign countries. His statue in Cambridge bears the inscription " From farmers and friends in many lands." His place, as the owner of the leading flock of Southdowns, was taken not by his old rival the Duke of Richmond but by the Earl of Walsingham, whose flock was at Merton Hall, Thetford.

A group of Red Poll breeders, particularly Sir Edward Kerrison, Lord Sondes and Sir Willoughby Jones, now seem to have been determined to make known the merits of their breed, and at the three Shows in 1863-65 they practically monopolised the award list in the classes for ' Other established breeds.'

Among the winners in the classes for the newer breeds of sheep are the names of many of the pioneer breeders—in Lincolns Lynn of Stroxton, Marshall of Branston and Wright of Nocton Heath ; in Oxford Downs Gillett of Bampton, Wallis of Bampton, Bryan of Witney, while outside the early home there were already the flocks of Henry Overman in Norfolk and of Charles Howard, of implement fame, at Biddenham by Bedford. With Hampshires Humfrey was still the leader, but he had strong competition from King of Lambourne, Lawrence of Wilton and Canning of Devizes. Breeders of Shropshires were much more numerous ; a considerable number of flocks was always represented in the prize list, and honours were much more widely distributed.

It was still impossible to recognise specific breeds of pigs other than the Berkshire, but some attempt at an improved classification of the remainder was made, separate classes being provided for pigs of " A Large White Breed," " A Small White Breed," " A Small Black Breed," and " Any other Breed." The Large White exhibits came chiefly from Yorkshire and Lincolnshire, and, although Herd-Books were still things of the distant future, the Large White Yorkshire pig may be said to have now taken fairly definite shape.

As recorded in Chapter VIII, the Show of 1866 was cancelled owing to the prevalence of rinderpest, and at that of 1867, at Bury St. Edmunds, the cattle exhibit was omitted as a measure of precaution. It is not often that the Society's Live Stock steward, in his official report, allows himself to indulge in whole-hearted condemnation, but Mr. C. Randell on this occasion did so. After drawing a contrast between past and present representatives of the Hampshire breed, and emphasising the remarkable progress that had been made in the preceding twenty years, he turns to rend the unfortunate Suffolk, then making its first bow before the general public.

The Black-faced Suffolk must possess qualities which a stranger knows not of. The men of Norfolk and Suffolk know their business too well to make it safe to assume that these sheep are as bad as they look, and there must be some merit, though not visible, to compensate for all their faults. Still, it seems difficult to understand that their good qualities, whatever they are, might not be retained with some modification at least of the long legs, short ribs, thin necks, bare backs and naked heads that characterise the Suffolk sheep shown at Bury.

Surely a cross with the Hampshire or Shropshire was indicated.

Fortunately the Suffolk breeders were not deterred by such criticism from their attempts to make a good sheep out of their so unpromising raw material, for the fashion for lean joints, which has led to the decline of some of the most popular breeds of those days, has been all in favour of the despised and rejected tribe of 1867.

The Leicester Show of 1868 was good but uneventful. That at Manchester in the following year created a new record, with an attendance of 200,000, a gate of £17,000 and a surplus of no less than £9,000. This success was partly accounted for by the fact that the Prince of Wales (Edward VII) was President of the Society, took a keen interest in the Show, and attended personally. Partly, too, the Society received more help from the local committee than on any previous occasion. The latter provided a prize fund for Scottish and North-Country breeds of live stock, and the collection of mountain sheep—including the Lonk,

Herdwick, Cheviot, Limestone, and Scotch Blackface—was easily the largest ever brought together.

Mr. W. Wells, in his report, mentions particularly the obvious utility of the deep-milking North-Country strain of shorthorn-type cows which were shown in one of the local classes, hints plainly that the Society might do more to encourage dairy qualities in breeds other than those of the Channel Islands, and adds a note on the question of human nutrition which has a modern ring :

It was surely most desirable, he says, to give every encouragement to these fine milk producers—a plentiful supply of milk for the infant Britisher being even more important than the supply of beef to the adult. Medical and other evidence showed clearly that the superior stature of the North-Country farm worker, over the town worker and the South-Country labourer, was due to the difference in the consumption of milk. " It would be better," adds Mr. Wells, " if the milking qualities of our three most prominent races of cattle were not, as is too often the case, overlooked in the desire to secure every additional pound of beef."

It is, in fact, a just criticism of English pedigree cattle breeders that, for so long, they neglected the dairy cow. Partly the explanation was that dairying long continued to be a small man's business, and that the working dairyman could not afford a long price for a sire ; partly, too, the breeders were bound to keep an eye on the export market, and the new countries were meanwhile not interested in milk production. At any rate England was not, for many years to come, to turn seriously to the business of improving the general run of dairy cattle. The Scots and the Dutchmen were allowed to steal a march.

Mr. Wells's other interesting comment is one of disquiet about jumping competitions, the first of which took place at Manchester under the auspices of the local committee.

Notwithstanding the financial success which attended this exhibition of horse jumping, it is very questionable whether it is advisable for the Royal Agricultural Society to allow the same thing another time. It was impossible to make the public understand that the Society had nothing to do with the proceedings.; and it leaves the Council open to the charge of encouraging, for the mere sake of money-making, what, after all, partakes more of the nature of a circus than of any part of an Agricultural Society's Show.

At Manchester the routine veterinary examination of the horses was discontinued as no longer necessary, though the Judges were instructed to call for veterinary advice if they were in doubt on any question of unsoundness. It seems that no disadvantage attended the new procedure, though it was agreed that its operation must be closely watched.

In 1870 the Society returned to Oxford after thirty-one years. It was now the hey-day of the corn-and-sheep farming of the Cotswolds and the chalk country, and the display of the South-Country breeds of sheep was the finest ever seen. There were large entries of Leicesters, Cotswolds, Lincolns and Ryelands and also of Southdowns, Oxford Downs, Shropshires, Hampshires and Dorset Horns. There was, too, the largest attendance of overseas buyers, of both cattle and sheep, that had ever been known, and the 'boom' in pedigree live stock, and more especially in Shorthorns, had well begun. In 1870 T. C. Booth of Warlaby sold two cattle for export for a thousand and fifteen hundred guineas respectively, and he is said to have refused a bid of two thousand for a seven-year-old cow; Captain Gunter sold a pair of heifers of the Bates "Duchess" family for 2,500 guineas. The demand came chiefly from the United States, Canada and Australia.

The name of Henry Dudding, who was later to become the best known breeder in England, appears for the first time in the Award List for this year; he took first and second prizes for Shearling Lincoln Rams, and also first for Shorthorn bull calves.

Jacob Wilson, who was already doing much for the Society, and who was later to do still more, was the senior steward of live stock at Wolverhampton in 1871. The weather was wet and the showyard a quagmire; some of the railways made exorbitant charges for the conveyance of stock, and there was dissatisfaction with the Council in their choice of the place of meeting. Nevertheless, there was a large and a successful Show, of which the most notable feature was a collection of more than five hundred Shropshire sheep. The most interesting part of Wilson's report is that in which he discusses the great responsibility of the Society's judges of the newer breeds such as the Shropshire. There were great numbers of good sheep among the representatives of this breed, but there was still a good deal of difference of opinion on the standard type. It was essential that the judges' decisions should be consistent in themselves, that they should be based on the same principles from year to year and that the judges should indicate, in their reports, the reasons which governed their decisions. The responsibility of the Society during the formative period of a breed has always been a heavy one, for the right guidance of the efforts towards a uniform and good type depends largely upon the wise selection of the 'Royal' judges.

In 1872 the country meeting was held at Cardiff—the first meeting in

Wales. It suffered by reason of a widespread outbreak of foot-and-mouth disease, and narrowly escaped another danger, for it opened on July 15 and the second visitation of rinderpest began two days later. The local Radnor sheep made a favourable impression on the visitors, but otherwise there was little of special note. The only innovation in the following year, at Hull, was the provision by a private donor of a prize fund for donkeys and mules. The Council considered the continuation of such classes, but came to the conclusion that it would be doing a better service to devote any available funds to the further encouragement of horse-breeding.

The prosperity of farming reached its climax in the early 'seventies and with prosperity came perhaps the greatest " crazy period " in the history of pedigree stockbreeding. The breed that was specially singled out was the Shorthorn, and the strains that were in special demand were not the Booth but the Bates ; indeed, it was the Bates " Duchess " tribe that came to be regarded as the *crème de la crème* of cattle.

English breeders travelled to the United States for the sale of Campbell's herd at New York Mills, where the purest of Bates blood was to be found, and, among others, two of the Duchesses were brought back at 8,120 and 7,000 guineas respectively. At the sale of the Dunmore herd, in Stirlingshire, in 1875, a sum of £26,000 was realised for thirty-nine head. Torr's Aylesby herd (of Booth blood), which contained some fine cattle, was sold in 1875, when 84 head averaged £510. Within the next ten years four more sales achieved averages of over £500. The affair was more like a wild stock-exchange gamble than a part of the sober business of stock-breeding.

At Bedford in 1874 there was a very successful Show, and the steward makes special comment on the steady improvement of the Red Polls. Taunton in 1875 would have been good, but was almost drowned out.

The three shows of 1876 to 1878 at Birmingham, Liverpool and Bristol respectively, were highly successful in every respect. Each left a substantial margin over expenses, and the Society's invested funds benefited to the extent of nearly £15,000 in all. Entries of live stock were large, and the Society was able considerably to extend its prize list. Classes were now provided annually for " Agricultural " (Shire), Clydesdale and Suffolk horses ; for Shorthorn, Hereford, Devon, Sussex, Longhorn, Jersey and Guernsey cattle ; for Leicester, Cotswold, Lincoln, Oxford, Southdown, Shropshire and Hampshire sheep ; and for Large White, Small White, Small Black and Berkshire pigs. At each meeting other breeds which had any considerable local importance were catered for. The area required for the Show had now increased to about seventy acres.

The prosperity of arable farming was a little on the wane, for serious competition from the new countries was beginning to tell upon corn prices ; moreover, 1877 and 1878 were poor seasons, when yields were down ; but the development of the new countries was creating a growing demand for British stock and, so far as the pedigree breeder was concerned, all seemed well in the best of worlds.

But these three years of prosperity were the prelude to the most disastrous venture that the Society had so far undertaken. The meeting of 1879 was due to be held in the London area, and the Council decided to celebrate the Society's fortieth anniversary by the holding of a Show upon an unprecedented scale. The Prince of Wales was President. A site of 100 acres was secured at Kilburn, on what was then the western edge of London. The prize sheet provided for every distinct British breed of horses, cattle, sheep and pigs, and also for goats, asses and mules. Moreover, provision was made for all the important breeds of horses, cattle and sheep of the adjacent continental countries—Spain, Portugal, France, Holland, Germany and Denmark. The total amount of prizes offered, for live stock and farm produce, was £13,200.

The Show was planned to remain open for seven days—from Monday, June 30, till Monday, July 7. There was an exhibit of ancient and modern farm implements, a working dairy illustrating both British and Continental methods of manufacture, and many other special features. The total of entries of horses, cattle, sheep and pigs was 2,874, or nearly double the previous record.

There was some anxiety about the nature of the soil of Kilburn Park, but what should have proved ample precautions were taken. A sum of over £3,000 was spent on preliminary draining and levelling, and 14,000 railway sleepers, 24,000 hurdles and 2,800 cubic yards of ballast were used in the making of roads and pathways. But the weather of 1879 broke all previous records. The ground was sodden throughout the time when the exhibits were being assembled, and the week of the meeting was marked by persistent rain and cold winds.

As the day for opening drew near, and the heavy machinery began to come in without the hoped-for change in the weather, it became painfully apparent that the struggle to get things in their places would be desperate and protracted. By the end of the week preceding the opening, the approaches to and the spaces between the implement shedding, as well as much of the rest of the ground, was worked by the incessant traffic into a wide sea of tenacious mud, through which it was impossible, without quadrupled horse-power, to move the heavy exhibits into their places.

The efforts made by the railway companies to . . . deliver the exhibits were unceasing. Horses in numbers were lent by other railways and gangs of men, relieving each other, worked through the night as well as day.

After the opening of the Show, as one wet day succeeded another, it was determined, as a last chance of success, to defer the closing of the implement department till Thursday, July 11. But the pitiless rain continued.

Queen Victoria gallantly carried out her promise to visit the Show, though her coachman did not dare to leave the sleeper-laid roads ; the Prince of Wales braved the weather on four separate days and did all he could to maintain the dampened spirits of the assembly.

Apart from the sad effects of the weather, the only disappointing feature of the Show was the poor representation of foreign cattle, the necessary quarantine regulations having proved a deterrent to many intending exhibitors. The attendance reached the considerable total of 187,000, but the revenue fell short, by £15,000, of the sum of £50,000 which had been spent on the venture. The loss represented nearly half of the Society's carefully accumulated funds.

The Kilburn show was but one episode in the worst year that any farmer could remember. Hay crops were ruined, wheat yielded less than half a normal harvest of miserable blighted grain, and stock wandered all summer on sodden pastures ; Professor Wrightson in a survey of the year, wrote : " I shall not readily forget the feeling of thankfulness with which I regarded twelve o'clock at midnight on December 31st, 1879. At any rate, a doleful ruinous year had departed." Farmers consoled themselves, as best they could, with the thought that such a season could not soon come again. But there were other troubles in store, unconnected with the English climate. The Golden Age of English farming had reached its end.

CHAPTER V

LIVE STOCK AND SHOWS, 1880–1905

IT might have been hoped that the God of Rain, after his onslaught at Kilburn, would have spared the Society for a time, but he returned to the attack in the following year at Carlisle. The Secretary of the local committee was a meteorologist and, according to his observations, 2·78 inches of rain fell in the yard during the period of the Show. There was anxiety lest the Eden and the Calder, near whose confluence the showyard was situated, should overflow and cause a general inundation ; this, fortunately, did not happen, but some of the parades had to be cancelled because it seemed unlikely that the animals would be able to complete the course. The North-Country men, however, proved to be of a hardier breed than the Londoners. We are told that they " streamed " into the ground in great numbers, determined to enjoy the show in spite of the elements. In the event there was an attendance of more than 90,000, and the deficit that fell on the Society's General Funds was no more than a few hundreds of pounds.

Finlay Dun, the distinguished veterinarian, in his report on the live stock, made some general observations and drew some useful general lessons. The farmers of Cumberland and Westmorland were already finding a recipe for depression ; they were giving up the attempt to grow corn in competition with the new countries, and were concentrating their efforts on better stock-farming. This meant, upon the one hand, better pastures and heavier root crops and, upon the other, the improvement of the local breeds of animals—Clydesdale horses, Shorthorn cattle and Cheviot and Leicester sheep. The progress made since the previous Carlisle Show was most impressive. There were now over a hundred herds of pedigree Shorthorns in the two counties, and half the prize-money for the breed went to local cattle. Moreover, the plain farmer was making good use of pedigree bulls, and the general run of Cumberland Shorthorns were already becoming widely known for their wealth of flesh, their useful milking qualities and their robust constitutions. Improvement had been no less striking in the case of sheep ; flocks of fine Cheviots had replaced most of the old nondescript stock on the hill grazings, and Leicesters and Border Leicesters were flourishing on the improved low-ground pastures.

For a good many years before 1880 the general organisation of the Show had been steadily improving, and had now reached a high degree of perfection. The task had been begun by Sir Brandreth Gibbs and was carried on with still greater energy by Jacob Wilson, who succeeded Gibbs, in 1875, as Honorary Director of Country Meetings.

Jacob Wilson, the son of a Northumberland farmer, was born in 1836. He was one of the early students of the Royal Agricultural College and one of the first two to be awarded the College Diploma. A little later he was one of the first small group of candidates in the Highland Society's examination, and was the only one, in the first year, to be awarded the Fellowship. After some years of farming with his father, he was appointed, in 1866, agent for the Chillingham estates of the Earl of Tankerville. From this time onwards he built up a very large practice as land agent, valuer and arbitrator. He was also, for some time, secretary of the Northumberland Agricultural Society and achieved remarkable success in running the County Show.

Thus, by 1875, Wilson had served a useful apprenticeship for the Honorary Directorship, and events were to prove that no happier choice could have been made. It was only his energy and resource, his indomitable spirits and his power of inspiring his staff, that saved Kilburn from complete chaos—" he never once lost his temper and never relaxed his exertions." In the space of a few years his unfailing tact and good humour won the goodwill of exhibitors, his clear-headed planning made for efficiency and precision in the showyard arrangements, and he was continually considering possible improvements. He made the Show more instructive to the farmer and, without pandering to the crowd that looked for mere amusement, he also made it more attractive to townsfolk. Wilson's term of office lasted till 1892 ; thirteen years later, at the end of his life, he returned to his old duties in an effort to save something from the wreck of Park Royal.

In the years that followed Kilburn it was an important object to rebuild the Society's reserve funds, and hence to avoid deficits on the Shows. To this end, again, it was necessary to spend with due regard to economy, and also to ensure a " gate." Wilson won his first notable success at Derby in 1881. His committee dispensed with a showyard contractor and did the work by direct labour, thus effecting a saving of

£1,300. There was a good display both of stock and machinery, there was an efficient and enthusiastic local committee and, above all, there was a week of perfect weather. The outcome was an attendance of 128,000 and a surplus of £4,500. Reading in 1882 practically cleared expenses while York (1883) and Shrewsbury (1884) both left substantial profits. Preston in 1885 added nearly £2,000, and the Society's balance sheet, by the end of that year, showed reserves of over £37,000. In other words, the financial loss of Kilburn had been more than made good. The Jubilee was now, however, drawing near, and a conservative financial policy was still necessary if that occasion was to be celebrated in a fitting manner.

A useful development of the 'eighties was the growing tendency of other bodies to co-operate with the Society in making each successive Show an event of special interest and value to the area where it happened to be held—for example, by providing extended prize lists for the locally important breeds of live stock. Changes and developments in agriculture are necessarily slow, and unless special efforts of this kind are made, there must be a tendency for each annual exhibition to be rather like that last. It was one of the advantages of the migratory system that it provided an opportunity to give variety to the Royal Shows by illustrating the diversity of British farming. Help was obtained from private individuals, from County Agricultural Associations, from Breed Societies, and, notably, from the specially constituted local committees. Thus at Shrewsbury, in 1884, the combined liberality of Mr. W. G. Webb, the Shropshire and West Midland Agricultural Society and the Hereford Cattle Society enabled a great extension of the prize lists for Shropshire sheep and Hereford cattle, and produced a remarkable display. Assistance of a similar kind continued to be given in other areas.

Another matter which received the Society's consideration at this time was the mode of selection of the judges of live stock. The qualifications required in a judge are both technical and personal—he must be an acknowledged expert, and be familiar with the generally accepted standards for his breed ; but he must also be prepared to make honest decisions without fear or favour. In 1884 the Council announced that it would be prepared to receive nominations from the various Breed Societies, and from its own members, and that all names submitted would be placed upon a list for the consideration of the Judges Selection Committee.

We note various developments and incidents in the live stock Show during the 'eighties. Lancashire had been the pioneer county in developing commercial poultry-farming, and hence it was felt that the Preston Show would be incomplete without a poultry section. This was provided —for the first time since Bury St. Edmunds in 1867. It was so successful that the Council decided to re-establish it as a permanent section of the Show. Dairying was the other outstandingly important branch of Lancashire farming, and hence there were many classes for non-pedigree dairy cows and for dairy produce.

About this time, a difficulty arose in connexion with the classification of draft horses. The " English Cart Horse Society " had been formed in 1878 and the name, four years later, was changed to the " Shire Horse Society." Up till 1882 all heavy horses, other than Clydesdales and Suffolks, had been grouped together as " Agricultural Horses," but the formation of the Breed Society changed the situation ; it was one of the aims of the ' Royal ' to encourage pedigree breeding, yet it seemed difficult to debar from competition animals that were " not in the Book," at least until it should be seen whether the Breed Society would succeed in amalgamating, into a real breed, the various and rather diverse local types of heavy horses. In 1883 special classes were provided for Shire stallions, and others for animals ineligible to compete as Shires, Clydesdales or Suffolks. This arrangement was repeated in 1884, but the non-pedigree classes were poorly filled and the quality was unimpressive. Consequently, in 1885, only one set of classes was provided for " Shire or Agricultural " horses ; but now the supporters of the " Book " were inclined to complain ; if one or other pedigree register was now open to all heavy horses of good breeding, why should non-pedigree animals be recognised at all ? In the event the difficulty largely solved itself, for at Preston only three non-pedigree stallions appeared, and none of these was adjudged worthy of a prize. The Shire Breed was now, in fact, soundly established, though provision was made, at some further shows, for unregistered agricultural horses.

In the Preston report we note that Suffolk sheep had increased and multiplied in the eighteen years since Mr. Randell made his disparaging remarks on those which appeared at the Bury St. Edmunds Show. Mr. Turner, to whom it now fell to report on the live stock, went out of his way to praise the breed, ". . . a variety of much merit, as should now be recognised by a separate class, combining as they do so large a quantity of mutton of fine quality with a fleece of more than medium weight, and being also extremely valuable for purposes of cross breeding." These words may be said to date the beginning of a widespread change in the sheep industry ; the popularity, with the consumer, of the big Longwool breeds was beginning to wane, and the smaller and leaner-fleshed Downs were beginning to rise in favour.

Norwich, in 1886, was, with the exception of Kilburn, the largest live stock Show that the Society had got together, the total entries of cattle, sheep, horses and pigs amounting to 1,825. The parades of the local breeds, particularly of Red Poll cattle, were the subject of much favourable comment.

Newcastle in 1887 was noteworthy as marking the beginning of energetic measures, on the part of the Society, to encourage light-horse breeding. We are already forgetting how large was the importance of the light horse in the England of fifty years ago. The large farmer had indeed, some years earlier, ceased to drive his carriage and pair ; but there were fast-increasing numbers of carriages in the towns, horse-buses and horse-trams were multiplying, and there was a large demand for army remounts. The home output had for some time proved unequal to these demands, and considerable numbers of animals, many of them of poor type, were being imported. In view of this position the Council voted a sum of £1,000 for the provision of premiums to be awarded for sires " suitable for getting hunters and other half-bred horses." This sum was offered at Newcastle in the form of five £200 premiums, payment of which was subject to certain stipulations regarding service fees and the number of mares covered. There was a good entry of 38 stallions, five excellent animals were selected and these were allotted by ballot to five districts in the northern show area. This was the first considerable effort to encourage light-horse breeding in England.

Apart from this special feature, the Newcastle meeting produced the finest collection of stock that the reporter, James Macdonald, had seen in the dozen years of his Royal Show experience. " There is not in the United Kingdom a variety of live stock of importance (if we except perhaps the once-noted but now declining Longhorn) that was not represented on Newcastle Moor." In particular the Scottish breeds were more fully represented than they had ever been at an English show.

In this year the Society suffered a severe loss by the death of its Secretary, H. M. Jenkins. Jenkins had carried out his secretarial work with unfailing efficiency, and had co-operated most loyally with Jacob Wilson in the development of the Show ; but his really distinguished and original work had been done as Editor of the *Journal*, many of his own contributions to which were of great value. His special interest was in the agricultural development of foreign countries, and on this subject he had become the acknowledged authority. He was succeeded, as Secretary and Editor, by Ernest Clarke (later Sir Ernest Clarke).

The Nottingham meeting of 1888 created, so far as live stock was concerned, another record for an 'ordinary' Show, and attracted the largest attendance since Kilburn. Since 1886 there had been a further development in connexion with light-horse breeding; the Royal Commission on Horse Breeding had reported in 1887, and had recommended steps similar to those already taken by the 'Royal,' but upon a much larger scale. On its recommendation, the Government provided an annual grant of £5,000 for distribution as premiums to suitable sires. A combined stallion show, arranged by the Society, was held in Nottingham in the February preceding the meeting, and from the 105 horses that paraded 27 were selected for premiums—22 for Queen's Premiums and the remaining five for those offered by the Society. This arrangement of a joint show, in the late winter, was continued for a number of years, the premium winners being sometimes brought to the summer Show to be paraded.

For the Society's Jubilee year Queen Victoria accepted the office of President, and appointed the Prince of Wales (Edward VII) as her Deputy. The Prince had already shown his keen interest in agriculture and in the Society (he had been a Governor for twenty-five years and had twice been President), and he now entered upon his duties with enthusiasm. He occupied the Chair at several Council meetings and on March 26, the actual anniversary of the granting of the Society's Charter, he entertained the Trustees, Council and Officers of the Society to a Royal Banquet in St. James's Palace. The occasion of the Jubilee was used to increase the Society's membership, and an accession of 2,425 new members brought the total to 10,984, or nearly 2,000 more than the previous highest figure. The ordinary rotation of country meetings was suspended, and the Queen gave her permission for the Show to be held in Windsor Great Park. Preparations were made upon a grand scale, provision being made for every recognised breed of stock in the country. Entries both of stock and implements exceeded expectations, and the area of ground originally allocated—97 acres—proved to be inadequate. It was extended, with the Queen's permission, by a further 30 acres, but, in the end, the implement space had to be rationed.

The live-stock entries were more than double those at Nottingham in the previous year, and largely exceeded those at Kilburn, the figures being:

	Windsor 1889	Nottingham 1888	Kilburn 1879
Horses . . .	996	546	815
Cattle . . .	1,644	644	1,007
Sheep . . .	1,109	537	841
Pigs . . .	265	148	211
Total .	4,014	1,875	2,874
Poultry . . .	861	343	—

The muster of cattle breeds consisted of Shorthorn, Hereford, Devon, Sussex, Longhorn, Welsh, Red Polled, Aberdeen Angus, Galloway, Highland, Ayrshire, Jersey, Guernsey, Kerry and Dexter—in other words, the only notable difference between the list of 1889 and the familiar one of a modern ' Royal ' is the absence from the former of the Dairy Shorthorn and Friesian. There were, indeed, classes for " Any Other Breed " in which a few Dutch cattle appeared ; and, in the prize lists of the milk yield and butter classes, both Shorthorn and Dutch cows appear.

The sheep comprised Leicester, Border Leicester, Cotswold, Lincoln, Oxford Down, Shropshire, Southdown, Hampshire Down, Suffolk, Dorset Horn, Kent or Romney Marsh, Devon Longwool, Ryeland, Dartmoor, Exmoor, Wensleydale, Roscommon, Limestone, Cheviot, Blackface Mountain, Herdwick, Lonk, and Welsh Mountain. In the class for " Any Other Breed " the prize-winners, except for a single pen of Merinos, were all South Devons.

The recognised breeds of pigs were the Large White, Middle White (first admitted to the list in 1882), Small White, Berkshire and Tamworth, and there was a class for " Any Other Black Breed " ; the latter produced specimens of both the Suffolk and the Devonshire strains, which were later to be amalgamated as the Large Black breed.

Thus the list, taken as a whole, was vastly different from that at the Society's early Shows, and not vastly different from that of the present day.

The prize fund benefited largely by contributions from Breed Societies, while the Local Committee gave £1,000, and a Mansion House Fund of £4,000 was produced by the City of London. The Queen gave twenty-nine gold medals as champion prizes for the more numerously represented breeds of horses and cattle. The total prize fund exceeded £11,000, in addition to £1,200 allotted to dairy produce, corn, wool, cider, honey, etc.

The Windsor Show was open to the public for seven days. It was favoured by excellent weather and the arrangements for the comfort and convenience of visitors were admirable. Queen Victoria returned from Balmoral for the event, and paid four separate visits to the yard, displaying, for a ruling monarch of fifty-three years' standing, astonishing energy and interest. The Prince of Wales was even more assiduous in his attendance.

All things considered, the Show was a great success—the only disappointing feature being, again, the rather moderate degree of interest displayed in it by the public of London. The total attendance, at 156,000, was large, but only some twenty-five per cent. more than that at Nottingham in the previous year.

The expenditure had been heavy, and it was not to be expected that it would be covered by the receipts ; the debit balance of £5,000 was, in fact, no more than had been originally anticipated, though the combination of a brilliant setting, brilliant weather and active Royal patronage might have been expected to attract a bigger attendance. The Queen, at the end of the meeting, expressed her satisfaction in a manner which gave the greatest pleasure to the Society ; Jacob Wilson was given a knighthood.

The next few years passed with but few important incidents. At Plymouth in 1890 the Shire Horse judges were not satisfied about the soundness of some of the exhibits that were placed before them, and asked for a general veterinary inspection. The actual number of disqualifications was small, but one or two of the animals which were turned down might probably, had they been allowed to compete, have taken high places in the prize lists. The incident, therefore, caused some disquiet. In the following winter the Council made a new regulation—or rather revived an old one—to the effect that no stallion should be awarded a prize until a veterinary inspection had shown it to be free from hereditary disease. In the following year the regulation was made applicable to mares as well.

By this time a large number of Stud and Herd Books had been established, and it was obviously desirable that the ' Royal ' should assist the Breed Societies in their aim of maintaining the purity of their respective breeds. It was, however, impossible to insist, from the moment of the formation of a Breed Society, that all animals of the breed, to be eligible for exhibition, should be officially registered. The line which had been taken was that, after a Breed Register had been in existence for seven years, any animal accepted for exhibition at the Royal Show must be entered in, or be eligible for entry in, the appropriate Stud or Herd Book. In order to ensure strict compliance with this rule, a further regulation was now made, requiring the exhibitor to produce proof that his animal was recognised, by the Breed Society in question, as eligible for registration.

In 1892 Sir Jacob Wilson retired from the Honorary Directorship, and his great services were gratefully acknowledged ; he was elected a

Life Governor and presented with a handsome piece of plate. His office was taken over by the Hon. Cecil Parker, but it was, in fact, impossible to fill Wilson's shoes. Some of the burden which he had assumed had now, perforce, to revert to the Secretary, and this, together with the increased membership of the Society and the growth of its activities, made it necessary to separate the offices of Secretary and Editor. The new Editor was Dr. William Fream.

In the same year a committee was appointed to review the scheme for the rotation of Shows, and it decided to include, besides the existing Show areas, the cities of London and Birmingham, and to hold the Show at Manchester and Liverpool in alternate cycles. The object of the new arrangement was to ensure that smaller towns should not be debarred from consideration, as Show sites, by reason of the fact that there was a large city in the same Show area.

From about 1890 onwards the old premises of the Society, at 12 Hanover Square, were felt to be inadequate to its growing needs, and various means of providing the necessary accommodation were considered. The view was also held by various members of the House Committee that it would be advantageous if the offices of a number of agricultural bodies, more particularly those of the Breed Societies, could be gathered together under one roof. If the 'Royal' could acquire a sufficiently large block of property, could let off portions of it to other bodies, and could provide a council chamber for general use, it was felt that it would be doing a useful service. In 1893 Harewood House, which adjoined No. 12 Hanover Square, came upon the market and seemed to provide an excellent solution of the problem. There was no time for the Society to act, but the Duke of Westminster and Sir Walter Gilbey bought the property privately, and held it until the Council should have time to consider its suitability. It was generally agreed that the opportunity was one that should not be missed, especially since the purchasers offered to hand on the property to the Society upon very favourable terms. Even so, the scheme was over-ambitious. The immediate cost was £37,000, and it was clear that a further large sum would be necessary to erect the large block of offices which the Committee had in mind. However, the purchase was completed, the purchase money being raised by the issue of debentures. A further £2,100 was provided by special donations, and this sum was set aside as the nucleus of a sinking fund. Thus, on paper, the Society's General Reserve Fund was left intact. The Council decided against the erection of a new building, which would have cost some £25,000, and in favour of the adaptation of the old house. The cost of this adaptation was very generously borne by the Duke of Westminster and Sir Walter Gilbey, and there can be no doubt that their liberality enabled the Society to provide itself with premises that it could not otherwise have afforded. Nevertheless, and in spite of the fact that some office accommodation was let off to the Shire Horse Society, the financial effect of the transaction was to increase the annual expenditure, under the head of rent and maintenance of premises, from about £800 to nearly £2,000 per annum. Moreover, this was not the only increase under the head of general administrative expenses, which rose from £3,300 in 1884 to £5,700 in 1895. The disaster which befell the Society in the early years of the twentieth century is commonly supposed to have been the result of the mistake of establishing the permanent showyard at Park Royal ; but the origin of the trouble lay farther back, in the over-ambitious financial policy of the 'nineties.

During the early 'nineties, however, the Show continued to prosper, though there was, in some quarters, the feeling that it was becoming a dangerously large venture. Doncaster in 1891 and Chester in 1893 were especially successful from every point of view. In 1894 the Show returned to Cambridge after the lapse of fifty-four years, and the Society received a most cordial welcome from both the University and the municipal authorities. The University had lately established a Department of Agriculture, and its Chancellor, the Duke of Devonshire, was also the Society's President for the year. The University celebrated the occasion by conferring Honorary Degrees upon several distinguished members of the Council, and on several officials of the Society. These included Sir Nigel Kingscote and Albert Pell (Doctors of Law), Lawes and Gilbert (Doctors of Science) and Dr. Voelcker and Ernest Clarke (Masters of Arts). The sixty-four acres of Midsummer Common made too small an area for a Show of the size that had now become usual, and entries had to be restricted ; nevertheless, the meeting was again highly successful. The attendance, considering the rather sparse populaton of the district, was remarkably large at 111,000. The only unfortunate circumstance was the absence of a pig section ; this had to be cancelled owing to a widespread outbreak of swine fever. The section had again to be abandoned at Darlington in the following year.

In February 1894 there died Sir Harry Verney, at the age of ninety-two. He was the last survivor of those who had met at the Freemasons' Tavern in 1838, and had long been the Father of the Society. He continued, until nearly the end of his life, to take an active interest in its affairs.

The Leicester Show of 1896 and that at Manchester in 1897 were again both large and successful ; the former attracted an attendance of 146,000 and left a profit of £3,600 ; the latter had a record attendance of 217,000 and produced a credit balance of £4,700. The financial condition of the Society was not, however, as good as these figures would lead us to suppose. Its regular annual income, from members' subscriptions, exceeded £10,000, but its normal expenditure reached about the same amount and its reserves were slender. Rent and administrative expenses amounted to nearly £6,000, the *Journal* (now a quarterly) was costing some £2,500, the cost of the laboratory and the salaries of the scientific staff amounted to about £1,500 and the net cost of running the examinations was nearly £500. In 1897, after two highly profitable shows, the Reserve Fund stood at £23,000, but a considerable part of this had been used to purchase Harewood House debentures, and the liquid assets amounted to only £15,000. This, in view of the rising expenditure on the Show, and the necessarily speculative nature of the business of running it, was far too small a nest-egg. The cost of the Manchester Show, for instance, exceeded £25,000.

The Council considered the financial situation, but failed to grapple with the problem, which was essentially one of cutting its coat according to its cloth. They decided, for the future, to charge against the Show a proportion of the Society's general administrative costs ; but this was no more than a book-keeping transaction, and did nothing to ensure against the risk of financial losses on the Show. Moreover, the long continuance of agricultural depression was now beginning to have an adverse effect upon the membership. This had risen from six thousand in 1875 to eleven thousand in 1891, but by 1898 there were signs that a decline was setting in.

In fact, a run of financially unsuccessful Shows now set in. Birmingham in 1898 should, according to past experience, have made a substantial contribution to the coffers, but it had poor weather, and an attendance of 98,000 was now not enough to cover expenses. The loss at £1,500 (after charging £900 for administration) was not serious in itself, but was none the less alarming. At Maidstone, in 1899, the Society spent £21,000 and lost £6,000. York, in 1900, was perhaps more disappointing, for it was a traditionally good centre and had fine weather ; yet it resulted in a deficit of £3,500. At December 1900 the Society's Reserve still stood in its Balance Sheet at nearly £19,000, but in fact its only liquid assets were £10,000 of Consols, and against this was a bank loan of £5,000.

Before York, and indeed immediately after Maidstone, it became clear that some reorganisation was necessary, and the Council appointed a Special Committee " to consider and report as to any alterations in

VIIIA.—A LETTER FROM QUEEN VICTORIA

VIIIB.—THE BUTTER-MAKING COMPETITION, WINDSOR, 1889

IXa.—Shire Stallion, " Harold." Born 1881
First Prize, Newcastle-on-Tyne, 1887

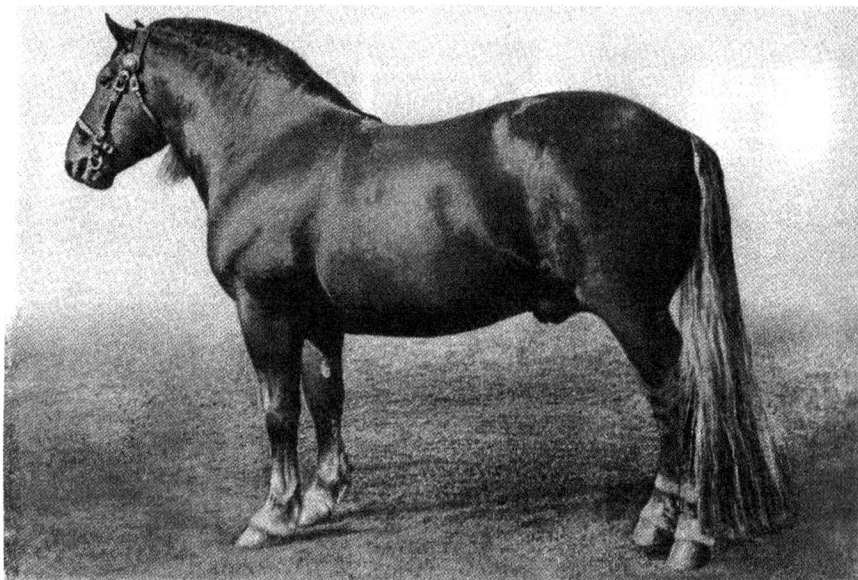

IXb.—Suffolk Stallion, " Wedgewood "
Winner of the Queen's Gold Medal, Windsor, 1889

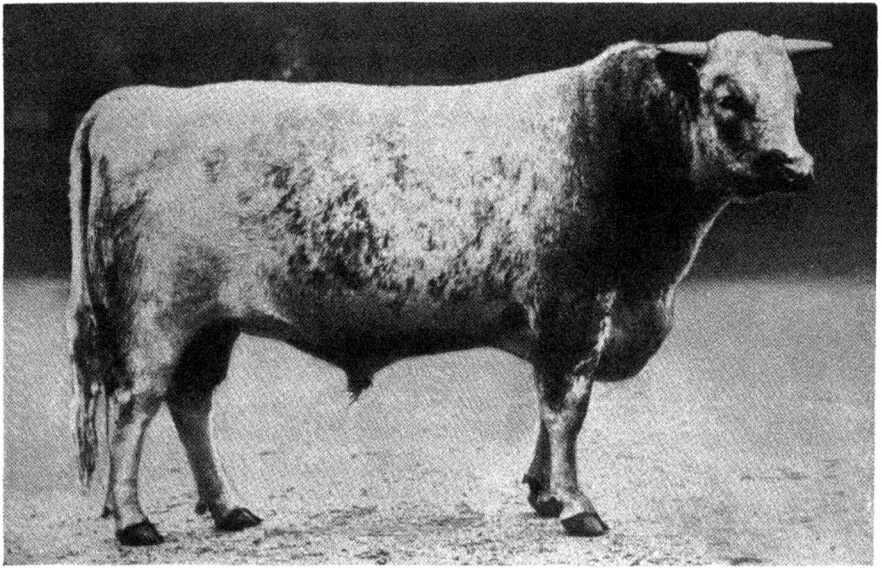

Xa.—Queen Victoria's Shorthorn Bull, "New Year's Gift"
First Prize (as a yearling), Windsor, 1889

Xb.—Hereford Heifer, "Primrose"
Breed Champion, Smithfield Show, 1888

XIA.—ABERDEEN-ANGUS COW, "WATERSIDE MATILDA II"
Winner of the Queen's Gold Medal, Windsor, 1889

XIB.—RED POLL COW, "EMBLEM"
Winner of the Queen's Gold Medal, Windsor, 1889

XIIIA.—Kent or Romney Marsh Ram
First Prize, Windsor, 1889

XIIIB.—Oxford Down Ram, " Liverpool Freeland "
First Prize, Preston, 1885

the present show system " such alterations to take effect after the current rotation should be completed in 1902. The Committee consisted of the Honorary Director (Mr. Percy Crutchley), the chairmen of the six standing committees which were directly concerned with the Show, and three unofficial members of Council. It reported to Council in the following February.

The report began by setting out past history ; it said that the migratory show system had been adopted at a time when a large agricultural show was a great event in any district ; but that other large agricultural shows had since been established and that these were now, in a real sense, competing with the ' Royal.'

The Society's Show had, upon the whole, continued to meet expenses until quite recently, but its steadily increasing size had meant steadily increasing costs, and there could be no assurance that the higher costs would be balanced by higher receipts ; indeed, the expectation must be that ' gates ' would get smaller.

Another difficulty was that of finding show sites ; the Show now required a block of nearly a hundred acres of level, well-drained land, preferably under old sward, with convenient access by rail. Such sites were scarce, and large expenses had lately been incurred in preparatory levelling and grassing down.

The financial success of a large and costly show depended upon the attendance of large numbers of the non-agricultural public. It had never, of course, been the primary object of the Society to make money out of its country meetings, but, unless these could cover expenses, the other useful activities of the Society would be imperilled. It was clearly necessary that the Society should husband its resources and, in the general interest of its members, limit the financial risks involved in the Show.

It had been suggested that the Show might be reduced in size, both by limiting the space allowed to implement firms and merchants and also by reducing the number of classes of live stock ; but, in fact, the Council was continually being pressed to provide more space and more classes. It seemed, therefore, that the way to the desired end was to reduce the cost of the Show without restricting its size.

The net financial result of the whole series of shows had been a balance of loss of £33,624, which had fallen upon the Society's general funds ; moreover, this result would have been worse but for the large contributions from local committees, amounting in all to £38,000

The only large item of expenditure that could be avoided was the cost of the annual erection and subsequent dismantling of the vast array of shedding and stands. " In view of all these circumstances, the Committee feel that they cannot take the responsibility of advising the Council to commit the Society to the holding of another series of shows on the basis of the existing rotation, which would involve the shifting of the Show from one district to another for the next nine or ten years."

" Taking into consideration all the facts of the case, the Committee have arrived at the conclusion that, if the Society's shows are to fulfil their proper function without an unwarrantable drain upon the Society's general resources, it would be desirable that they should be held upon a permanent location (preferably in the centre of England) which would be convenient for railway access from all parts of the country. In fact, the endeavour of the Society in the future should be to bring the people to the Show, and not the Show to the people."

The site would be equipped with permanent roads and buildings, with a railway station and with facilities for handling heavy machinery. No doubt fewer excursionists would attend on the shilling day, but " the regular visitors will come to the Show wherever it may happen to be held." An additional advantage would be that the site could be let for other purposes at periods when it was not in use by the Society. No suggestion was made on the question of the actual site, that question being left over until the Council should decide upon the principle of a permanent showyard.

The report was signed by all the members of the committee, but Sir Jacob Wilson made the following reservation :

" I have signed the report subject to the reservation that I am not satisfied that every alternative to the proposed permanent showyard has been exhausted. I am of opinion that, by various rearrangements, the size of the showyard could be appreciably diminished and that, by varying the composition of the prize sheet according to the wants and circumstances of each district visited, the annual expenditure on prizes and for the preparations consequent thereon could be much decreased."

The Committee's report came before a special meeting of Council on March 7, 1900, and was fully debated.

Lord Derby pointed to the rapidly improving facilities for travel and to the success which, in Canada, had attended the system of fixed shows. Jacob Wilson, having had time for reflection, was now more than doubtful about the proposal, and spoke strongly against it. He pointed out that the balance of loss on the Shows had been heavy and consistent only in the first eighteen years of the Society's operations and that, over the subsequent forty-two years, there was a balance of profit, despite the very heavy loss on Kilburn. He suggested that the Council were attaching too much importance to the results of the past two years. He felt that a Show taking place annually on the same site would be out of reach of the vast majority of small farmers and of practically all farm workers ; and he believed that the proposed plan would have a very bad effect on the recruitment of members. He freely accepted some blame for the present unwieldy size of the Show, but he submitted that it would be easy to reduce its size— for instance by arranging a rotation of years for the exhibit of the less important breeds of stock. With a fixed show the Society could look for no local support, and the lack of this would mean a loss of £2,000 a year. He thought there was, among the ordinary members of the Society, a general feeling against the Committee's proposal. The time might come when the change proposed would be for the Society's good, but he was convinced that the time was not yet. " He would implore the Council to pause for a while before abandoning a well-tried system which had worked so well for sixty years . . . in favour of a most doubtful and dangerous experiment."

The Earl of Feversham made a plea for consultation with the general body of members. Mr. D. P. Foster supported this, and joined with Sir Jacob Wilson in his profound misgivings. Mr. Martin Sutton said it would be in the interests of trade firms such as his own to have quarters in a permanent showyard ; but he feared that the Council, in this matter, was out of touch with the general opinion of members. His own impression was that the farmers were against the proposal, and he urged that a plebiscite be taken.

After further discussion Sir Nigel Kingscote, the Chairman of the Finance Committee, expressed his fears of financial disaster if the Society continued with its migratory show, and he pointed out that members would have an opportunity of expressing their views when the report of Council was presented at the Anniversary Meeting in May. The question was then put to the vote. The Committee's report was adopted by 38 votes to 4, and it was reappointed, with instructions to make enquiries and submit proposals with regard to possible sites.

In August the Committee reported again. It was now their view that the showyard should be in London rather than in the Midlands, and they wished to have the opinion of the Council on this point before proceeding with detailed enquiries.

Wilson returned to the attack. He had looked back over the statistics of past shows in the London area, and he did not think the prospects of a permanent London Show could be described as encouraging. He thought that, with a fixed site in the Midlands, the ' Royal ' would lose its particular character, but that it might probably remain a real agricultural show. With a London site the Show would more probably cease to be agricultural in any real sense. It would degenerate into some kind of horse-show for the benefit of professional showmen, and would lose its hold upon the general body of stockbreeders.

Mr. Martin Sutton asked whether the Council had not already voted in favour of a site in Central England ; whether the Committee had definitely reversed their previous opinion on this point ; and whether the opinion of members might not be taken—firstly, for or against the fixed site, and secondly (in the event of a majority in favour) for a London site or, alternatively, one near the geographical centre of the country. Albert Pell pointed to the many counter-attractions, including other shows, with which a ' Royal ' in London would have to compete, and to the improbability that the rank and file of the industry would come to London.

Sir Nigel Kingscote replied that all these considerations had been fully argued in Committee, and that he himself was fully convinced in favour of London. Sir Walter Gilbey spoke more strongly in the same sense.

Mr. Joseph Martin took the line that the Society had been placed in its present dilemma chiefly from financial reasons, and suggested that the whole subject of the Society's financial administration should be reviewed ; but nobody seems to have thought the point worth taking up.

Anyone reading through the debate to-day must get the impression that the balance of argument lay against London ; but the weight of official opinion was on the other side, and the voting was 34 for, and 12 against, acceptance of the Committee's recommendation. Accordingly the Committee began its search for a London site, and the Council proceeded with the arrangements for what were to have been its last two country meetings—at Cardiff in 1901 and Carlisle in 1902. The first made a profit of nearly £2,000, but Carlisle maintained its reputation for wet shows, and there was a loss of nearly £3,000.

Meanwhile, in May 1901, the Special Committee reported that, after prolonged exploration, they had found a site admirably adapted to the Society's needs. This was at Twyford Abbey, between Willesden and Ealing, and extended to 102 acres. The Council accordingly took a lease for 50 years at an annual rental of £1,000, with an option to purchase, on defined terms, both the leased land and an additional 15 acres. The Council were now faced with the problem of raising the very large sum of money needed to equip the showyard. It was obvious that the Society's

XIVa.—Park Royal Show, 1904

XIVb.—Liverpool Show, 1910

financial position would not enable it to borrow anything like the amount required, and therefore it was decided to open a subscription fund. By December of that year nearly £28,000 had been raised. This was sufficient to warrant the Council's completion of the purchase of the land (for £26,146), but a good deal short of the sum required to provide even temporary showyard equipment, which was all that was envisaged at that stage. The cost of the land and the estimated cost of the temporary equipment were together £42,000. The method adopted for the administration of the property was to form a private company under the name of Park Royal Limited, the share capital of £15,000 being held by the Society and the remainder of the money required being raised by the issue of £22,000 of debentures.

The rest of the story of Park Royal may be told in a few lines. Three Shows were held, in 1903, 1904 and 1905, and the figures may be allowed to speak for themselves :

	Live Stock Entries	Attendance	Expenditure	Receipts	Balance (Loss)
1903	2,212	65,013	£28,301	£18,621	£9,680
1904	2,056	52,930	£21,395	£14,475	£6,920
1905	2,161	23,978	£18,577	£11,298	£7,279

Total loss £23,879

After the 1904 Show it was decided that the Society had no alternative but to cut its losses and get out of the venture. It was then, however, too late to arrange for a country Show in 1905 and, after some hesitation, it was decided to carry on at Park Royal for one more year. Jacob Wilson and his supporters had, from the time that the decision was made, loyally co-operated in the effort to make Park Royal a success, and for the third Show he returned to his old post as Honorary Director. He rallied exhibitors to his side, and produced perhaps as fine a Show as any of those of his hey-day. " Throughout the week he worked as few besides himself could work, being here, there and everywhere, inspecting, supervising and arranging the details from day to day. He personally conducted the King round the showyard and performed the same duty for the Prince of Wales and other distinguished visitors." But it was all in vain ; his public failed to come. Three days after the Park Royal closed its gates he was taken ill, and in three more days he was dead.

CHAPTER VI

LIVE-STOCK IMPROVEMENT AND SHOWS, 1906-38

IT is not easy, even at this distance of time, to write with assurance about the Park Royal venture—to say just how unwise it was, or to analyse the causes of its failure. Some things are certain, some questions are still arguable, and, of course, it is possible only to speculate about the likely outcome of the alternatives which were discussed by the Council and were advocated by the minority party.

Among the certainties is that the Society's financial difficulties began a good many years before the proposal for a fixed showyard was first mooted. The running of the Show had always been a speculative undertaking and the Society, in its earlier years, had very properly made provision for losses to be met out of its calculable income from members' subscriptions. A reserve fund was gradually accumulated, and this had been more than adequate to meet the heavy deficit on the Kilburn Show of 1879. A few years of careful and conservative finance sufficed to make good this loss. From about 1885 onwards there was, however, a slipping away from the old sound and conservative policy. Show profits were too often treated as current income, while Show losses had too often to be met from reserves. This is clear from the following statement, which sets out, in round figures, (a) the total reserve fund, or surplus capital as appearing in the balance sheets of 1885, 1895, and 1900 ; (b) the value of the gilt-edged securities, representing the easily realisable or liquid assets, in each of these years, and (c) the expenditure on the Show of each of the years in question.

	Total Reserve Funds £	Liquid Capital (Gilt-edged Securities) £	Expenditure on Show £
1885	38,000	31,000	17,000
1895	25,000	15,000	18,000
1900	19,000	10,000	21,000

The insurance fund was thus being allowed to fall, while the financial risks were steadily increasing.

Again, it is highly probable, if not quite certain, that a plebiscite of members, such as was suggested at both of the crucial Council meetings, would have shown, in the first case, a majority against a permanent showyard, and, in the second, a preference for a site elsewhere than in London. It is true that members had the opportunity of a General Meeting to veto the Council's proposals ; but members had long been accustomed to entrust the Council with all broad questions of policy, and they were given little time to organise the opposition which seems to have been rather generally felt. The Council would seem to have been, as was said by several of its own members, out of touch with opinion in the Society at large.

Thirdly, whatever the financial outcome had been, one can hardly escape the feeling that the Society's educational work would have suffered if it had settled its Show permanently in London. It was too much to expect the working farmers or farm workers of the north or west country to make the journey to London at what is normally a busy season on the land. Larger numbers would possibly have gone to Derby or Leicester, which were other suggested sites ; but how many more it is impossible to estimate with any certainty.

Lastly, and perhaps most important, Park Royal was at the best a heavy gamble, and the Society could ill afford to lose its stake.

When the outcome of the 1904 Show was reckoned up it became a question whether the Society could any longer carry on its accustomed activities. Its only assets, apart from fixtures and plant, consisted of £13,000 in Harewood House Debentures and the shares in Park Royal ; these latter had cost £27,000, but their value was now problematical. An appeal was made for a guarantee fund for the third (1905) Park Royal Show, and more than £7,000 was subscribed ; had the sum been much less the Show could not have been held at all.

The Council now saw that a complete reorganisation of the Society's affairs was essential, and they began by devising a new method for the election of the Council itself. In theory, election had always been in the hands of the general body of members, for it was made by a General Meeting ; but the Council nominated for all vacancies, and in practice their nominations had almost invariably been accepted. The new plan, which was laid down in a Supplementary Charter, gave the members in each district the absolute power of electing their Council representatives. The Supplementary Charter was sealed in the spring of 1905, the first election under the new rules took place in July and the new Council, containing a large proportion of new men, held its first meeting on

August 1st. Mr. F. S. W. Cornwallis (afterwards Lord Cornwallis) was the first President under the new régime. There was a general discussion on the Society's position and prospects, after which it was decided :

" That a Special Committee be appointed, with power to call for any information from officials and to obtain professional assistance, if they think necessary, from accountants, solicitors and valuers, to thoroughly investigate the entire position of the Society and to make a report to the Council . . . as to what reforms and economies they consider desirable to put the Society on a sound footing."

The report of the Committee, which was presented on October 4th, was as follows :

1. In accordance with the Resolution of the Council of August 1, 1905, your Committee met on September 20 and 21, and beg to report on the reforms and economies they consider desirable to place the Society on a sound footing.

2. The Committee desire to express their recognition of the services of the staff in the past ; but in view of the financial position of the Society they are unable to recommend a continuance of so large an expenditure as is at present incurred under this head, and regret that they see no other course open to them but to request the Council to ask for the resignation of the whole Staff as at present engaged at Hanover Square and Park Royal.

3. They consider that a sum not exceeding £1,500 per annum is all that the Society is at present justified in expending on the salaries of the Secretary and administrative staff.

4. They recommend that a Secretary (to devote his whole time to the work of the Society) should be appointed at a salary of £600 per annum, and an assistant at £300 per annum.

5. HAREWOOD HOUSE.—They recommend that if a satisfactory price can be obtained for Harewood House it shall be sold at as early a date as possible, and failing this, that such part of the house not absolutely required for the purposes of the Society shall be let.

6. JOURNAL.—They recommend that the cost of the *Journal*, including distribution, shall not exceed £600 per annum.

7. SCIENTIFIC DEPARTMENTS.—They recommend that the Board of Agriculture be approached with a view to obtain a grant in aid of the scientific operations of the Society, which must otherwise be curtailed, unless such assistance can be obtained.

8. THE SOCIETY'S SHOW.—They recommend that no Show be held at Park Royal in 1906, but that it take place in the provinces if a suitable site can be obtained and financial arrangements made ; also that a sum of not less than £2,000 from the Governors and Members' subscriptions be credited to the expenses of the annual Show.

9. They think it would be most advantageous that a conference be held

annually between the Council and Officers of the Royal Agricultural Society and the Secretaries of the County, Breed and other leading Agricultural Societies to consider questions of general and mutual interest.

10. Park Royal.—They recommend that immediate steps be taken for the disposal of the Society's interest in the Park Royal Estate.

11. Subscriptions.—The Committee desire to draw attention to and emphasise the desirability of encouraging Members to give financial support to the Society beyond the minimum subscription.

Thomas L. Aveling
Richardson Carr
R. P. Cooper
R. Forrest
John Gilmour, Bt.
Gilbert Greenall
W. Harrison
Chris. Middleton
T. S. Minton
Fredk. Reynard
John Rowell
R. Stratton
G. Taylor
John Thornton
C. W. Wilson

(The Earl of Jersey and Mr. E. W. Stanyforth were unavoidably absent).

This report was adopted except for paragraph 10, which was remitted to the Committee for further consideration. Some members thought that the Park Royal property might rapidly appreciate in value, and that the Society might do better by holding on than by forcing a sale. The vendors had a right of pre-emption, at the original purchase price, until 1911, and it was thought that the sale might perhaps be postponed, with advantage, until this right had expired.

The Committee duly reconsidered the matter but, at the November meeting, they reiterated their view that the property should be sold forthwith ; they had come to the conclusion that the cost of maintenance, together with the interest charges, would probably be greater than the capital appreciation. The vendors having declined to exercise their right of pre-emption, the estate was put in the market and was sold, with its fixtures, for £28,500. Harewood House was vacated, in order to facilitate its sale, in September 1906, and the headquarters were moved to 16 Bedford Square. Early in 1907 Harewood House was sold for

£45,000, or £8,000 more than it had cost. There was opposition to both these steps by sections of the Council, but Mr. Adeane had the strong support of men like Sir Richard Cooper, Sir Gilbert Greenall, Richardson Carr, Thomas Aveling and William Harrison, and, with their help, was able to convince the opposition.

As regards the other recommendations of the Committee, the employment of all the members of the office staff was terminated. Sir Ernest Clarke was given a substantial gratuity, and was made an Honorary Member of the Society. Even so, this action was an extremely disagreeable one for the Council to take, for Sir Ernest Clarke was a distinguished man who had served the Society faithfully for eighteen years. But the Council held the view that, since they had to make a fresh start in a much more humble way, it was better to start with new men. The new secretary was Mr. Thomas McRow.

In December 1905 a new Honorary Director was appointed in the person of Sir Gilbert Greenall (Lord Daresbury). He was destined to preside over the Country Meetings for a period of twenty-five years and was a worthy successor to Sir Jacob Wilson. Lord Daresbury was a breeder of many kinds of live stock, and few men who have attempted so much in this line of work can have won so large a measure of success. He bred Hackneys and Hackney ponies, Thoroughbreds, Dairy Shorthorn cattle and Large White pigs. His pig herd, especially, won so outstanding a position that he must be counted among the greatest improvers of the breed. He was also, by general consent, the best all-round judge of horses of his time in England. Mounted on his white cob, he became a universally known figure in the Showyard, where his quiet efficiency exerted a very valuable influence.

He had no wish to introduce revolutionary changes in the Show, but its revival from 1906 onwards owed a great deal to him. He was specially valuable to the Society in securing the closer collaboration of the Breed Societies ; probably nobody has presided over so many of these as Lord Daresbury did, and he was no less popular in those circles than in the ' Royal.'

His greatest success was in 1925 at his own county town of Chester. Here he occupied the three offices of President of the Society, Honorary Director of the Show and Chairman of the Executive of the Local Committee. The subscriptions to the Local Fund were the highest in the Society's history, and the Show is remembered by many old showgoers as perhaps the pleasantest as well as one of the finest they have ever attended.

1925 was to have been the last year of Lord Daresbury's Directorship but, at the urgent request of the Council, he continued to serve for four years more. At the General Meeting of 1930 he received the most cordial thanks of the Society for his work and was given a piece of plate as a memento of his period of office. His task was taken over by Mr. U. Roland Burke (Sir Roland Burke), who completed the century and whose greatest achievement was the Centenary Show (see Chapter XIV).

We must now return to 1905 and trace the process by which the finances of the Society were rehabilitated.

In that year the Society decided to revert immediately to its old migratory system for the Show, and received an invitation to go to Derby in 1906. Sir Richard Cooper, a member of Council, very generously undertook to meet any deficit and the Local Committee, the County Agricultural Society and the Breed Societies all rallied round the hard-pressed ' Royal.' The result was an excellent exhibit, a good attendance and a surplus of some £2,000.

Meanwhile Mr. Adeane had succeeded Sir Nigel Kingscote as Chairman of Finance, and under him the Committee set about its difficult task. Mr. Adeane's first review of the financial position, in January 1906, was necessarily tentative, for the Society's only substantial assets were the properties which the Council had resolved to sell, and it was impossible to say what they might realise ; but it seemed that they might bring in about enough to pay off the debt, leaving no large balance one way or the other.

The actual outcome was better than had been anticipated, largely through the success achieved in the disposal of the real property (which produced a surplus of £11,000), and partly because of donations by members. The recovery was greatly helped by the success of the first three Shows after Park Royal, for Derby's £2,000 was followed by surpluses of £5,000 at Lincoln in 1907 and of fully £10,000 at Newcastle in the following year. Meanwhile the general account had been made to balance by the introduction of a strict system whereby every spending committee was " rationed " and required to submit an annual estimate. By the end of 1908 the available capital stood in the Balance Sheet at more than £43,000, and the immediate crisis had been definitely surmounted. The Finance Committee did not, however, consider that its task was done, but advised that the Society should continue to build up the Reserve Fund. The scheme that they devised was to set aside part of the general income (at first £2,000 and later £3,500 annually) as a contribution to the Show account, and to add all Show surpluses to the Reserve.

The subsequent financial history of the Society has been entirely happy. The Reserve Fund reached £50,000 by 1910, withstood the strain of the difficult post-War period, and had risen to £160,000 by 1930 and to £246,000 by 1938. The improvement was partly due to a rise in the value of investments. In recent years the interest on its invested funds has amounted to about half the Society's total income, and latterly the 'Royal' has not only been able to make adequate provision for all the activities that fall within its particular sphere, but has been able to make several substantial contributions to research stations and educational institutions.

The debt of gratitude which the Society owed to its Chairman of Finance was acknowledged by Lord Mildmay when, at the July Meeting of Council in the centenary year, it fell to him to make reference to Mr. Adeane's retirement. After touching upon Mr. Adeane's Presidency of 1917, and his valuable work as Chairman of the War Emergency Committee and Treasurer of the Relief of the Allies Fund (see Chapter XIII), Lord Mildmay said that Mr. Adeane's outstanding service had been in watching over their finances. He had taken over control when the Society was almost bankrupt, and it was due to him that they had reached their present position, with reserve funds that were ample for any eventuality that could be foreseen. Mr. Adeane's careful economy and his wise and judicious investment policy had saved the 'Royal,' and he moved "That the most grateful thanks of the Council and of the Society be conveyed to Mr. Charles Adeane for his invaluable services as Chairman of the Finance Committee, extending over a period of thirty-three years."

In other respects, too, the leave-taking from Park Royal marked the beginning of happier times. The weight of depression had begun to lift from the farming industry as a whole ; stock-breeding was almost flourishing, dairying was expanding, and some breeds of live stock were enjoying a boom in the export trade.

Notable among these was the Lincoln sheep, which naturally turned out in great force when the 'Royal' visited its home county. The greatest of the Lincoln breeders, and probably the greatest breeder of his generation in England, was Henry Dudding, the tenant farmer of Riby Grove, near Grimsby, who was an active member of the 'Royal' Council. His farm and his flock were very large, his Shorthorn cattle were only less notable than his sheep, and his annual sale became one of the great events of the

LORD CORNWALLIS
1864–1935

CHARLES ADEANE, C.B.
CHAIRMAN OF FINANCE, 1906–39,
PRESIDENT, 1917

XVIa.—Lincoln Ram. Champion at Derby, 1906
Shown by Henry Dudding and afterwards sold for 1,450 guineas

XVIb.—Suffolk Ram, " Playford Model "
First Prize, Park Royal, 1904

agricultural year. In 1906 his draft of 56 shearling rams sold at an average price of £151, while the eight sheep that he had taken to the Derby Show realised an average of £563 and his Derby Champion (see Plate XVIA) was sold to Argentina for 1,450 guineas—the British record price for a sheep. In the same year his 51 Shorthorns averaged £99, with a top price of 1,000 guineas.

The export trade in British pedigree live stock has been important, from various points of view, over a long period of years. It has provided the foundation for the development of the live-stock industry in the Dominions, the United States, Argentina and the new countries generally ; it has conferred great benefits upon the British consumer by providing the source of abundant imports of high-quality meats and dairy produce ; and it has provided valuable income to home breeders, especially in times when other branches of the industry have been depressed. On occasion, however, a difference between the requirements of the exporter and those of the home farmer has placed the pedigree breeder in a dilemma, and such a dilemma faced the Shorthorn men in the years about the beginning of the present century.

The Shorthorn had become pre-eminent among our breeds of cattle in the days of the Collings, when it was quite definitely a dual-purpose breed. There was some divergence between the more fleshy Booth and the dual-purpose Bates strains and, in the latter half of the nineteenth century, Scottish breeders, under the leadership of Cruickshank, produced a highly specialised beef type. This was the kind of Shorthorn that was wanted to stock the ranges of the United States and the estancias of Argentina, and was that which commanded the highest prices from about 1890 onwards. Up till this time the North-of-England breeders had maintained and improved their strain of Shorthorns upon the old dual-purpose lines, but many were now tempted to ' top ' their herds with Scotch bulls, in order to obtain the thick-fleshed beasts which made the overseas buyer loosen his purse-strings. Finally, a small minority of South-countrymen still shared the view expressed by Richard Stratton, who wrote in 1897, " The dairymen of this country should be the best and most numerous customers for pure Shorthorns . . . if the foreigners fail us, as they probably will some day, where will be the outlet for our bulls ? " In 1901 the Shorthorn Society, largely by Stratton's persuasion, instituted prizes at certain county shows for " Pedigree Shorthorn Cows of Milking Characteristics."

Up till 1903 the Royal Agricultural Society provided only one set of classes for the breed, and the judges had often to deal with a variation of type that made any logical placing almost impossible. In 1905, however, the Dairy Shorthorn Association was formed,[1] and in that year the 'Royal' provided classes for younger and older Dairy Shorthorn cows. The number of exhibitors was at first small, for many owners of milking Shorthorns had been driven to the use of non-pedigree bulls ; but numbers rapidly grew. In 1911 a complete classification was provided (for bulls as well as cows), and since that date the Dairy Shorthorn exhibit has grown until it is now often the largest in the cattle section of the Show.

Another newcomer appeared at Norwich in 1911 in the British Friesian, then known as the British Holstein. This Dutch breed was the only one of non-British origin that had played any important part in the cattle industry of the new countries. It had been under systematic improvement in Holland, just as the Ayrshire had been in Scotland, during the time when the main interest of English pedigree cattle breeders had been centred on beef. If the Shorthorn men had 'agreed to differ' ten years earlier than they did, the Friesian might not have been required. As matters stood, the breed supplied a real need, for it was a reliable and heavy milk-producer well suited to the better sort of dairy land. The foundation stock for the British Friesian was not very satisfactory, having been obtained from the rather mixed Dutch cattle imported in the time (especially between 1889 and 1892) when our ports were open to live animals. The Breed Society (formed in 1909), however, obtained special permission from the Board of Agriculture to import a number of cattle from Holland and, in 1914, fifty-nine very carefully selected specimens were introduced. Subsequent importations and careful breeding led to steady improvement, and soon after the end of the Great War the foreign breed had established an important position in this country.

Another development of the Show at this period was the introduction of milk-yield and butter classes for the dairy breeds of cattle. It has always been a criticism of live-stock shows in general, as a means of encouraging live-stock improvement, that there is a tendency in the show-ring to place undue emphasis upon various quite unimportant characteristics of the animal—on such points as colour pattern or the shape of horns. Few breeds have entirely escaped harm from this over-emphasis of 'fancy' points as opposed to those which affect real utility. In the case of the

---

[1] The promoters of the Dairy Shorthorn Association were Sir Oswald Mosley, Lord Crewe, and Messrs. Charles Adeane, Richardson Carr, Walter Crosland, R. W. Hobbs, F. Punchard, C. A. Scott-Murray, Richard Stratton and George Taylor.

meat breeds, however, fashions and fancies tend to be short-lived—the butcher and the bacon curer know what they want, are shrewd judges of actual values, and their strictly business outlook soon brings the fancier back to earth.

In the case of dairy cattle the difficulty is much more real. The outward and visible signs of milking capacity are, at the best, somewhat doubtful and unreliable, and breeders who are concerned to produce profitable herds of cattle must rely more upon milk records than upon showyard performance. Some have even taken the view that the influence of the showyard, on the improvement of dairy stock, has been harmful. It would seem, however, that the dairy breeder must, in the long run, look at his cows as well as at their records ; otherwise it would be easy to ' improve ' the cow to such a pitch that life for her would be impossible under the conditions of the ordinary farm. If this be true, the showyard must remain a useful complement to the record book.

The Council realised in the early years of the present century that something should be done to encourage the breeding of cattle with the object of higher production, and their first step, in 1903, was to institute regular butter classes. This was eight years before any means for official milk-recording was provided in England. Two classes were provided for cattle—for those over and under 900 lb. weight respectively. The prizes were awarded, according to a scale of points, on the amount of butter actually churned from the milk produced in twenty-four hours by the cow ; one point was awarded for each ounce of butter and, in order to handicap the cattle fairly, one point was added for every completed ten days since calving, deducting the first forty days. The prizes in the first year were nearly all won by Jerseys. In 1905 milk-yield classes were instituted for cows of each of the recognised milking breeds, it being a condition that every cow entered should have been judged in the inspection class of its breed. The milk of each cow was weighed, sampled and analysed for butter-fat, and the order of merit was determined by a scale of points which took account of the quantity of milk, its fat-content and the stage of lactation of the cow.

The Jersey Cattle Society gave a good deal of help in supplementing the prize fund at a time when the Royal had little money to distribute.

It is well known that a one-day test is a much less satisfactory measure of milking capacity than a lactation record ; nevertheless, the milking trials were very useful in the days before official milk recording was begun, and they still serve to maintain the balance as between productive capacity and the other things that go to the making of a good cow.

Under Sir Gilbert Greenall's able direction the Show, from 1906 till 1911, pursued the even tenor of its way. At Doncaster in 1912, however, there was a very difficult situation to be dealt with. On the Friday of the week before the Show it became known that Foot and Mouth Disease had been confirmed in a lot of cattle which, on the preceding Monday, had been exposed for sale in Liverpool Market. The cattle from the market had not all been traced to their respective destinations, but it was feared that the infection might have been widely spread. Greenall made arrangements for a specially searching inspection of all the cattle, sheep and pigs as they arrived at the railway station and at the showyard. On the Sunday night the Board of Agriculture telegraphed that it might be necessary at any moment to impose a general " Stand-Still Order ", and exhibitors were given permission to remove their animals forthwith if they so desired. On Monday morning the Board made an order prohibiting the exhibition of cattle, sheep and pigs at the Show, and when the pressmen arrived for their private view they found the stock being loaded for their return journey. When the prohibition was received no less than fifteen hundred animals were already in the yard, and a further seven hundred were on the way. The Show staff and the railway company, however, co-operated closely and worked energetically, and all the cattle, sheep and pigs had been removed from the ground before the gates opened on Tuesday morning. Eight of the cattle were suspected of having been exposed to infection, and these had to be slaughtered ; otherwise exhibitors escaped without loss. The Show, despite this misfortune, drew a good attendance, while a good many exhibitors generously declined to reclaim their entry fees, and the loss, at £1,200, was unexpectedly small.

The Bristol and Shrewsbury meetings of 1913 and 1914 passed off well, and the Show continued to be held in each of the two earlier years of the Great War, though in 1916 the implement exhibit was greatly curtailed because many of the implement firms had turned over to munition work. Preparations were begun for a Show in 1917 in the hope that, by the summer, the country would be at peace ; but the project was soon abandoned. There was, naturally, no thought of a Show in 1918. During these years the Council became more and more actively engaged in war work, some account of which is given in Chapter XIII.

A start was made again in 1919 at Cardiff, where the Show had been due in 1917. With the general rise in prices the costs of the Shows rose, in the post-war years, to rather terrifying figures. Cardiff cost £40,000 and the outlay at Darlington in 1920 reached the unheard-of sum of £67,000. But farmers had money (many of them for the first time in their lives) and were ready to reinvest this in much-needed replacement of equipment. They attended the Shows in great numbers, and implement makers did a roaring trade. There was a roaring trade, too, in pedigree live stock, for American and other overseas farmers had large war profits to invest. Some indication of the extent of their demands may be obtained by comparing the pre-War with the post-War trade in Shorthorns. The following figures refer to the sales, at public auction, of Shorthorns in Scotland :

					Number Sold	Average Price
						£ s. d.
1912	.	.	.	.	1,326	46 0 5
1913	.	.	.	.	1,598	48 14 9
1914	.	.	.	.	1,426	51 18 8
1919	.	.	.	.	1,710	219 10 6
1920	.	.	.	.	2,235	254 2 0
1921	.	.	.	.	2,233	102 10 4

A few other examples may be quoted. William Duthie, the leading Beef Shorthorn breeder of his time, sold his year's output of bull calves in October 1920, at the astonishing average price of £1,400. In the same year top prices of 1,100 guineas were obtained both for Lincoln and Border Leicester rams, and a Gloucester Spots pig realised over £600.

In some of the post-War years the Show produced unheard-of profits for the Society, the record figure of over £19,000 being reached at Newcastle in 1923 ; fortunately these profits were conserved, for other times were to come. The other extreme was Newport, in 1927, when the debit balance reached £10,800.

It is an old criticism of this country's agriculture that, while it possesses some of the finest live stock in the world and has played the leading rôle in supplying other countries with high-class breeding material, a large proportion of its animals, and especially of its cattle, should still be indifferent or bad. The Royal Agricultural Society has played an important part in improvement at the top end but, by itself, it could do nothing towards the elimination of the unfit. In this matter Northern Ireland set an example when, in 1922, the Government passed the Live Stock Breeding Act. This required the licensing, by the State, of bulls of the prescribed age and the branding, as rejected, of all that were regarded as unfit to beget good stock. In 1925 a Bill on similar lines was drafted by the Ministry of Agriculture, and the Council expressed its full approval of the principle which it embodied. The Bill was dropped owing to the opposition of the National Farmers' Union.

In 1927 the Veterinary Committee of the Council returned to the question, and in May of that year Sir Merrik Burrell, its Chairman, strongly urged that the matter be not allowed to rest. Experience in Ireland went to show that its scheme was easily workable, that it was imposing no hardship upon breeders and that it was undoubtedly producing highly beneficial results. He supported this contention with a great deal of evidence. Lord Bledisloe said that there had been deep disappointment over the dropping of the Bill, that the whole of the Irish experience went to show that the fears of English farmers were unfounded, and that he hoped that opposition would soon be withdrawn. Sir Archibald Weigall quoted Australian experience as pointing to the same conclusion. The Council unanimously reaffirmed their former resolution in favour of the Licensing scheme. The campaign was pressed in other quarters by Sir Merrik Burrell and his supporters, and at last, in 1931, the Improvement of Live Stock (Licensing of Bulls) Act reached the Statute Book.

In 1930 a minor but rather difficult problem of show organisation arose out of the decision of the Ministry of Health to bring into force certain of the provisions of the Milk Designations Order 1923. The point was that, unless the organisers of shows were prepared to provide special and separate accommodation for cattle from licensed herds (Certified and Tuberculin Tested), the holders of licenses would be prohibited from exhibiting their cattle at shows. At first it seemed that it would be impossible, at large shows such as the 'Royal,' to ensure the measure of isolation that the Ministry required. But with some modification of the requirements and with every willingness on the part of the Society to do all that was possible, the problem was solved. The new arrangement of separate stalling for the different groups of cattle has meant some inconvenience to the cattle stewards and the exhibitors, but this was a small price to pay for the continuance of the competition of owners of licensed herds.

A development of the Show from 1930 onwards has been the growing provision for the events organised by the Federation of Young Farmers' Clubs. Since that year the international stock-judging contest has been held in the Showyard, and teams from Scotland, Northern Ireland, Wales, Canada, the United States and England have on some occasions competed. From 1935 onwards facilities have also been provided for a Young Farmers' Cattle Show, which has grown from year to year.

In 1933 the Society made an exception to its long-standing rule that none but pure-bred and pedigree live stock should be provided with classes. This took the form of special classes of commercial pigs. It was felt that, if the pig industry was to expand in face of the growing competition from overseas, it was urgently necessary to improve the quality and the degree of uniformity of home-produced bacon and pork, and it was believed that commercial classes would demonstrate to producers not only the types of pig required by the bacon curer and pork butcher but the pitch of condition to which they should be fed. Accordingly, classes were provided for both pure-bred and first-cross animals of both types, and the judging was entrusted to men familiar with the requirements of both trades. The classes were continued for three years, and undoubtedly served a useful purpose ; but with the introduction of payment according to grade, which was a feature of the Pigs Marketing Scheme, it was felt that the demonstration had been rendered unnecessary.

It is unnecessary to describe the features of recent Royal Shows, which will be familiar to most readers of this book. Briefly, with the exception of that of 1932, those of the 'thirties were uniformly successful. If we exclude special occasions such as Kilburn and the Jubilee and Centenary Meetings, those at Ipswich in 1934 and at Wolverhampton in 1937 were probably the largest and finest that the Society has produced. The Centenary Show of 1939 is dealt with in some detail in Chapter XIV.

CHAPTER VII

IMPLEMENTS AND MACHINES

THE means adopted by the Society, in its early years, for encouraging improvements in farm machinery, have been indicated in Chapter III. The plan was to draw up an elaborate classification of the many types of machines and to offer a substantial prize for the best of each, further money prizes and medals being provided for any new implement which the judges might think worthy of recognition. By 1845, as we have seen, the scheme was in difficulties in that the organisation was failing to keep pace with the rapid increase in the number of entries. At Newcastle in 1846 there was no trouble, for the number of exhibits (735) was substantially less than that of the previous year. But at Northampton in 1847 there was a renewed and unprecedented increase to 1,321 entries, and the judges must have had a very harassing task in selecting, from among so many, the winners of the thirty-two awards. Since it seemed that still further increases were likely, the Council decided on a thorough overhaul of the organisation. Ten judges were appointed—two engineers for the steam-engine class and eight farmers for the others ; the Friday, Saturday and Monday preceding the Show were set aside for trials and tests, and it was laid down that none but Judges and Officials were to be admitted to the yard until the Tuesday ; the Society's Engineer was to carry out any necessary engineering tests and to advise the judges, at their request, on points of construction ; finally the staff of stewards was increased.

The task of setting the revised scheme to work fell mainly upon two individuals, and its success was largely due to their joint efforts. By 1848 Josiah Parkes had so many irons in the fire that he felt obliged to resign his appointment as the Society's Engineer, and the work was given to the consulting firm of Easton and Amos of Southwark ; the Society's work was allocated to the junior partner, C. E. Amos, whose qualifications for it were supremely good. Amos had had very little schooling but, as a boy, had learnt farming from his father and mechanics from his grandfather, who owned a millwright's business at March in the Isle of Ely and who had a good deal to do with pumping plant and other drainage equipment in the Fens. Amos served an apprenticeship with another millwright, gained experience in various branches of engineering and, at the age of 31, joined Easton in practice. He was a man of real ability and ingenuity, was keenly interested in the farmer's problems and gave the Society his enthusiastic service over a period of twenty-three years.

From the outset he was most happily associated with Henry Stephen Thompson of Kirby Hall, Yorkshire (later Sir Harry S. Meysey Thompson), who was a wealthy landowner and, at the time, one of the most active members of the Council. Thompson was an enthusiast for all kinds of agricultural improvements and had been mainly responsible for the formation of the Yorkshire Agricultural Society in 1837. He was prominent in other spheres of activity ; under his Chairmanship the North-Eastern Railway Company became one of the largest and most successful business undertakings in Britain, and he also founded and guided the United Companies Railways Association, which performed the very necessary function of looking after the shareholders' interests during the period of railway speculation. He was to serve the Society in other capacities than that of Implement Steward ; in 1855 he succeeded Philip Pusey as Editor of the *Journal* and after 1859, when a salaried editor was appointed, he continued, as Chairman of the Committee, to direct the policy of the *Journal*. In the same year he entered Parliament. He was the Society's President for 1867 and remained active in its affairs until, in 1873, his health failed him. In spite of a shy and aloof manner he commanded the loyal service of all who worked under him, and his capacity for organisation was outstanding.

The scheme of 1848 remained in force for seven years, the Society providing classes for all varieties of implements and the judges applying scientific tests wherever possible. In 1855 the scheme was in fact modified by classifying the implements of husbandry into three groups and providing classes for only one of these groups in each year ; there is little doubt, however, that the 1848 plan was the best adapted to the circumstances of the time. Its objects were clearly set out by Thompson in his report on the York meeting in 1848, and were three. Firstly, the farmers in each of the Show districts must be given, in turn, the opportunity of seeing the whole range of modern machinery, and must have the guidance of the judges, and the results of carefully conducted tests, to help them in their choice of tools for their respective requirements. Secondly, the smaller local makers must be given an opportunity of seeing the work of the leading makers such as Howard, Ransomes, Hornsby, Garrett and Clayton, of making arrangements for the supply of such components as could best be turned out on a factory scale, and of making arrangements, through the payment of royalties, for the use of patented ideas. Thirdly, although the large makers, in 1839, would have laughed at the suggestion that they had anything to learn from the Judges, the position had now changed. Amos's dynamometers and other instruments could measure the draft of a scarifier, the man-power required to turn the handle of a chaff-cutter, the horse-power developed by a steam-engine or the power required to drive a thresher, and his results often upset the complacency of an exhibitor and set him to remedy unsuspected defects in his machine.

As we turn over the pages of the old implement reports we may see how interest concentrated for a period upon one or two types of machines and then, when these had reached a satisfactory level of efficiency, turned to something else. In the early 'forties drills were the main centre of interest, but by 1846 most of the inferior types had been eliminated, and there was little to choose between the models that continued to appear at the Shows. In 1848, when Thompson and Amos took control, the greatest excitement was being displayed about tile-making machines. The legislation of the period had made available many millions of money for the financing of land drainage—partly from public funds and partly private money provided through the agency of The Lands Improvement Company and other similar bodies. The urgent problem was to produce, at a cheap enough price, the vast numbers of tiles that were required.

Up till the 'thirties a great variety of materials—straw, brushwood, turf, peat, stones and bricks—were used to form the conduits of covered land drains. Hand-made tiles of the old horse-shoe pattern were the common solution in districts where good clay was available. The clay was " pugged " to obtain an even consistency and was then packed by the hands of the worker into a wooden frame, the top surface being planed off by means of a thin wire. The resulting slab of clay was turned out of the moulding frame and was either left flat to form a " sole " or was care-fully bent over a " horse " into the arched tile that formed the upper part of the drain. Tiles made by such a process were necessarily expensive. The first cylindrical pipes seem to have been made by John Reade, a gardener, in 1843, but his machine was crude and unsatisfactory. The first notable success was scored by Scragg who, as we saw in Chapter III, won the Society's prize in 1845.

A host of competitors now entered the field. " No class of agricultural implements," says Thompson in his report on the York Show in 1848, " has been so rapidly improved as drain-tile and pipe-making machines. Only five years have elapsed since the Society's meeting at Derby, at which but two were exhibited, and they excited so little attention that the bare enumeration of them in the Implement Report was thought sufficient ; whilst at York 34 were shown and it may safely be said that, with the exception perhaps of the steam-engines, no description of implement received more patient or ample trial." Whitehead of Preston won the prize, but another machine by Clayton, and an improved model by Scragg, were close runners-up. There was intensified competition at Norwich in 1849, though the actual number of machines was less, and Whitehead was again successful. His machine, worked by two men and a boy and turned by hand power, was set to make two-inch tiles 13½ inches long, and its output was 185 per hour.

The principle of all these early machines was the same. The clay was first pressed through a screen to remove stones, and was then passed to a pug-mill. The conditioned material was then fed into a cylinder and forced by means of a piston, actuated by a hand wheel, through a die, emerging as a long hollow tube. This was cut into equal lengths by means of a series of wires stretched on a frame, and the tiles were removed for drying and burning. A few years later larger machines, driven by horse or steam power, came into use.

The first of the portable steam-engines were contemporary with the earliest tile machines, but improvement was at a slower pace. The criteria on which the judgments were made were mechanical efficiency, simplicity of construction, durability, safety and cost. The routine efficiency tests determined the time, and the amount of fuel, required to raise steam, the brake horse-power and the fuel consumption per horse-power-hour. The Engineer Judges whom the Society appointed were able to give expert criticism and suggestions, which led to steady improvement. In 1848, for instance, their report stressed the importance of economy of fuel and indicated the principles upon which such economy depended. The prize engine of 1849 (Garrett) burned 11½ lb. of coal per horse-power-hour ; that of 1850 (by the same maker) 7½ lb. ; of 1852 (Hornsby) 4⅝ lb. ; and of 1853 (Clayton) 4⅓ lb. Concurrently the weight of the engines, in relation to their power, was reduced, and design was simplified in various respects.

The chief purpose of the portable engine was of course to drive the portable thresher, and attention was devoted to both concurrently. The thresher was an old invention, the first efficient machine having been invented by Meikle as far back as 1785. Old stationary engines, as well as horse-gears and water wheels, had been in general use on the large farms of the Lothians and Northumberland as early as 1820. The threshing-machine riots of 1830–31 give us the date of their widespread introduction in the South. But it took a good deal of ingenuity to combine the actual threshing with the complete dressing of the grain, and to arrange the whole mechanism in a box that should be small enough and light enough, when mounted upon four wheels, to be moved along country lanes. Economy of power, too, was necessary if the whole mechanism was to be driven by an engine of moderate size. Amos's equipment showed up the loss of power in some of the machines, and again indicated the way to improvement. The winning machine at Norwich in 1849 required 2·78 horse-power when running light, while at Exeter in the following year the corresponding figure was 1·39.

We may now note a tendency on the part of the big implement makers to specialise in certain types of machinery. Up till this time each of the large manufacturers, like Ransomes or Howard, aimed at providing for most, if not all, of the farmer's needs. But the range of requirements was now becoming too wide. The firm which specialised most completely on portable engines and threshers was Clayton and Shuttleworth, and the following statistics of its output of engines indicate the rapid expansion of the demand, which came partly from Britain and partly from overseas :

1851	140
1853	293
1855	491

Between 1849 and 1883 this firm sold over seventeen thousand threshers and nineteen thousand engines.

Interest in engines and threshers was maintained long after 1851, but in that year it was overshadowed, for the time being, by a sudden excitement over reaping machines. There had indeed been various attempts, from 1780 onwards, to invent a reaper ; in 1803 the Highland Society had awarded a prize to Gladstones, a Scottish millwright, for a machine which cut with a smooth-edged circular knife attached to a rotating wheel ; in 1812 Smith of Deanston (better known in connexion with land drainage) built a better machine, on the same rotary principle, which won a rather favourable report and a prize from an Agricultural Society in the Lothians. After various other attempts a definite step of progress was made by Patrick Bell, a Scottish parish Minister, who had a prize from the ' Highland ' in 1828, and from this time onwards a few Bell machines, turned out by local makers in Scotland, had actually been at work in Scotland each harvest. Bell's cutting mechanism, however, consisting of a series of pairs of scissors, was too complicated, and too liable to break down, for any but a mechanically-minded farmer.

XVIIa.—A Threshing Scene (About 1855)

XVIIb.—Bell's Reaper

XVIIc.—McCormick's Reaper

From the " Journal " for 1851

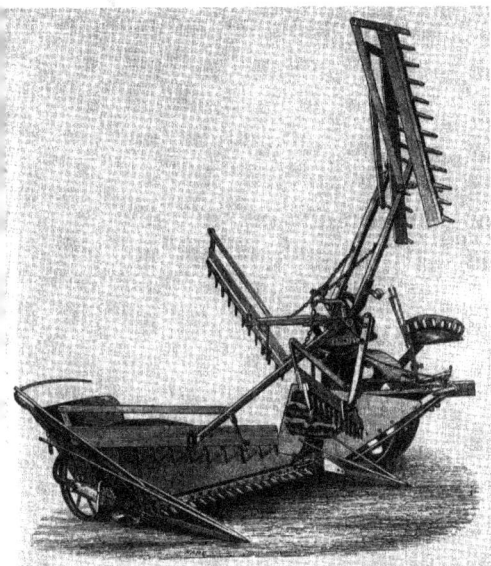

SIDE-DELIVERY REAPER (HORNSBY)
Winning Machine, Reaping Machine Trials, 1877

FOWLER'S PLOUGHING ENGINE

FOWLER'S DOUBLE-ENGINE STEAM PLOUGHING TACKLE
By Courtesy of Messrs. Fowler, Leeds

In 1851 there was held in London a great " Exhibition of the Works of Industry of all Nations " and a section was provided for Agricultural Implements. The Society decided to omit its own implement exhibit, and to hold its live-stock Show in the near neighbourhood of London, Windsor being the eventual choice. Philip Pusey was put in charge of the Agricultural Implement section of the International Exhibition, and his fully illustrated report on this, in Volume 12 of the *Journal*, is a very interesting and complete account of the state of agricultural engineering at the time. By far the most interesting of the machines on show, however, were the American reapers of McCormick and Hussey, the former of which, with Bell's machine, is shown in Plate XVII.

A trial, with unripe wheat, was arranged at Tiptree Hall, the farm of the illustrious John Joseph Mechi, and McCormick's machine was given the medal which the Royal Agricultural Society had offered. The following is Pusey's account :

The machine, drawn by two horses and carrying two men, a driver and a raker, cut the wheat about eight inches from the ground with the utmost regularity. The horses found the work light, though the machine was cutting at the rate of 1½ acres per hour. . . . The raker, standing behind the driver to rake the cut wheat from the platform, certainly had to exert himself ; but it is obvious that he and the driver, who has only to sit on the dicky, might very well exchange places from time to time. As one cannot put a high price on the labour of farm-horses at such a time, it is plain that a great saving must be effected by this machine, and every farmer can calculate it for himself, as he will also see the advantage of being rendered independent of the arrival of strangers to get in his corn. . . . This trial was witnessed by many farmers and no fault was found with the work. . . . Where ridges and water-furrows exist some difficulties seem to arise. But, on this level land, it was wonderful to see a new implement working so smoothly, so truly and in such a masterly manner. The fact is, however, that it is not an untried implement. Though new to this country, it has been used for some years in America, where experience has enabled the inventor to correct, in successive seasons, the defects invariably found in new implements. It is certainly strange that we should not have had it before, nor indeed should we have had it now but for the Great Exhibition, to whose Royal Originator (The Prince Consort) the English Farmer is clearly indebted for the most important addition to farming machinery that has been invented since the threshing-machine first took the place of the flail.

Further and more elaborate trials were arranged in connexion with the Lewes Meeting in 1852, and elsewhere in both England and Scotland. It was found that both the American machines required adaptation to the heavier crops grown in Britain, and that there were imperfections in both.

There was difference of opinion as between the McCormick and Hussey models, and at the Scottish trials an improved Bell machine was preferred to both. But the farmer had not awaited the results of these tests, for Thompson, in his report for 1852, says that within twelve months of the first trial " 1,400 machines have been ordered from the four leading makers."

In 1855 and again in 1856 and 1857 we find that the leading models were an improved McCormick, made by Burgess and Key (and embodying the new principle of an Archimedean-screw platform which delivered the cut corn in a continuous swath at the side) and an improved Bell machine manufactured by Crosskill of Beverley. The former was awarded the prize in the first and last of these three years, but the judges in 1856 preferred the other. By 1861, when trials were carried out at Leeds, there were three distinct types of machines, viz. manual and self-delivery reapers and combined reaper-mowers.

Meanwhile the steam-engine was attracting renewed interest, as it became clear that the difficulties in the way of its application to tillage were being rapidly surmounted. Under the triennial system of implement trials the first year of each cycle was to be devoted to implements of tillage, the second to sowing and harvesting machinery and the third to barn equipment. Chelmsford in 1856 was the first Show under the new system, and the Council had indicated its hopes for the future of steam culture by offering a prize of £500 for " the steam cultivator which shall in the most efficient manner turn over the soil and be an economical substitute for the plough or the spade." There were only two entries ; Mr. Smith of Woolston sent a cultivator which was worked by an engine and windlass placed in a corner of the field, but this was disqualified on the ground that it was a scarifier and did not invert the soil. The other was the first of John Fowler's steam ploughing outfits, a single-engine set working with a wire rope. The full trial was carried out at Mr. Fisher Hobbs' farm at Boxted Lodge, Essex, and was combined with the reaper trials which had been arranged, as an exception to the triennial plan, in the same year.

On a lovely day in the middle of August [says Clare Sewell Read, who was one of the judges] the reaping machines were started upon two splendid fields of wheat. The grain from one was so ripe that it was carted, threshed, ground into flour and served up at Mr. Fisher Hobbs' hospitable dinner-table that evening. Upon a portion of the cleared wheat-field Fowler's steam plough made capital work, riving up the dry Essex clay with great ease and completely inverting a well-cut furrow.

Sewell Read thought, in after years, that the judges might well have given Fowler the prize, but the carefully calculated cost of the work was 7s. 2½d. per acre, whereas it was possible to argue that it might have been done, with horses, for 7s. Fowler was disappointed, for the steam plough was the favourite child of his fertile brain ; but he was not discouraged ; he returned to the contest at Salisbury in 1857, and although he was adjudged superior to his three competitors, the cost of ploughing could still not be brought below the estimate of the previous year, and the prize was again withheld ; it was, however, offered again in connexion with the Chester Show in the following year. This time the judges were unanimous ; not only was Fowler's apparatus the best of those exhibited but it was beyond question " that Mr. Fowler's machine is able to turn over the soil in an efficient manner at a saving, as compared with horse labour, on light land of 20 to 25 per cent ; on heavy land of 25 to 30 per cent ; and in trenching of 80 to 85 per cent ; while the soil is left in a far more desirable condition and better adapted for all the purposes of husbandry."

Fowler's apparatus, in spite of growing competition, won the smaller premiums that were offered in 1859, 1860, and 1861. In 1862 there was a non-competitive demonstration in connexion with the great Battersea Show, and about a dozen sets, from eight different makers, were seen at work. The stewards reported that steam cultivation was " now becoming a great fact," but that a good deal more required to be done. Numbers of improvements had lessened the wear and tear on the ropes, anchors and pulleys, but the direct costs of doing the work had not materially lessened.

Meanwhile there appeared in the *Journal* numbers of lengthy articles in which every aspect of mechanical cultivation was discussed. John Algernon Clarke wrote a fifty-page prize essay on the subject which appeared in the *Journal* for 1859, and, five years later, he reconsidered the subject in the light of the more recent progress. In 1860 the editor at the time, Mr. P. H. Frere, collected and reviewed the experiences of many practical users. The Farmers' Club and other bodies gave the subject a good deal of consideration. It is impossible to summarise the mass of contemporary literature on the subject, but certain broad conclusions emerge. The early idea of using an engine as a tractor for drawing ploughs, cultivators and other tools, was now being abandoned ; the loss of power was greater than that arising from the use of a windlass and wire rope and, despite a good deal of ingenuity in spreading the weight, the damage to the soil from the passage of a powerful steam engine was serious. There were advantages, in the way of convenience and speed, in the two-engine system, but against these there was the heavier capital expenditure. Finally there was a strong belief among the more theoretically minded, a belief that was most forcibly expressed by Wren Hoskyns, that the fundamental idea of the leading inventors was ill conceived ; these, Hoskyns suggested, were obsessed with the idea that an engine should do the work of a given number of horses and do it in the same way—viz. by pulling such tools as ploughs and cultivators, which had been invented to make use of the tractive power of animals. Surely an engine, which produced rotary motion and could do so much more work, should be used in its own way—to invert and break down the soil in a single operation. Practical attempts to invent rotary tillers had in fact already been made ; as early as 1858 Thomas Ricketts of Birmingham had entered a machine for the £500 prize and, although the machine broke down in the trials, the Judges were satisfied that " the principle of rotary cultivation has taken a distinct position as a desirable and valuable addition to the mechanics of agriculture." Attempts to apply the principle have occupied many men's minds at intervals during the subsequent eighty years, but the rotary tiller is still far from having established itself as a standard farm implement.

By this time " steamers " were becoming a fairly familiar sight on large arable farms, and few of their owners seem to have regretted their investments. Many could show a saving as compared with their previous staffs of men and animals, and were able to earn extra income by contracting to plough and cultivate for their neighbours. It was on heavy land, which required three or four horses, or four oxen, for a single-furrow plough, that the gain was greatest, and since wheat stood at a very profitable level (the average from 1855 to 1874 was 54s. per quarter), farmers had every incentive to more intensive cultivation of the stronger soils, and profits were available for re-investment. Indeed, it was being argued that tile drainage and steam tillage were capable of doing for the heavy land what turnips, clover and sheep-folding had done for the light.

As an example we may quote the experience of Mr. E. Ruck, who farmed 800 acres of arable land near Cricklade, in the upper Thames Valley, and who read a paper to a meeting of the Society's Council in May 1863. He and his father had formerly worked the farm with fifty-six oxen (besides some horses)—seven four-ox teams for the morning shift and seven more for the afternoon. In 1859 he had bought a fourteen-horse-power steam-tackle set from Fowler and this had since done all the heavy tillage on the farm. He had pulled out five miles of hedges and thrown his farm into seventy-acre fields ; he had paid for his tackle very largely by the sale of his oxen, and he was satisfied that his tillage was at once better and more cheaply done.

The climax of the Society's efforts in connexion with steam tillage came in 1866, the year when the Show had to be abandoned on account of rinderpest. The report of the Council to the May meeting said :

The application of steam power to the cultivation of the soil has received the careful consideration of the Council, and they consider that the time has now arrived when an attempt should be made to arrive at the results which have been obtained by its use on different soils and in different localities. With this object they have appointed a Central Committee, composed of members of the Council for every district in England, to conduct a complete enquiry. Inspection Committees will visit such farms as the Central Committee may select for inspection. To assist them in their investigations paid secretaries will be attached to each Inspection Committee. A sum of £500 has been voted for the purpose of carrying out the investigation.

The actual cost of the investigation was nearly £700, apart from the expense of printing a report which, both in volume and in authority, might be compared with that of a Royal Commission. It occupied 330 pages in the *Journal* for 1867, set forth the advantages of steam tillage and discussed the various difficulties in the way of its general application. It showed that the double-engine system was speedier, and required fewer men, than the older single-engine system with its complicated arrangement of anchors and pulleys, and that steam tillage was of greater value upon heavy than upon light land. It set out, in terms of acreage, the capacities of the various sets of tackle that were available, and it indicated that the exploitation of steam by the small farmer lay in co-operative ownership or in the formation of steam-plough companies. Thompson, endeavouring to state the broad conclusion, said :

As a solution of a mechanical problem steam tillage is no doubt a most perfect and thorough success. That which was proposed has been accomplished, and great feats of cultivation have been performed which no other power could possibly have undertaken with the same result. But then comes the question whether, as a commercial speculation, steam cultivation has generally been a success ? I should be inclined to describe it as a success which a very small amount of ignorance and inattention would convert into a failure, a success where well managed, and a failure where badly managed.

It seems unnecessary to trace in any detail the latter-day history of steam tillage. There were many more trials, and a good many minor improvements ; there was sporadic interest and there were revivals of hope in rotary tillers and steam diggers ; in the late 'sixties and early 'seventies hopes of something like wholesale mechanisation ran high ; thus

IMPLEMENTS AND MACHINES

William Sanday, reporting on the record display of machinery at the Leicester Show in 1868 said :

The one prominent moral of the Show may be said to be this—that in a few years every operation of field tillage will be performed by steam. How this will be done is another question, but the general direction that will be taken is not very hard to predict. . . . It will best be taken up as a distinct speculation, and sets let out for hire.

The main development was in fact along the lines that Sanday predicted ; the man with a thousand acres of arable land might have his own tackle, but far more sets were bought by contractors. But the great expectation was never fulfilled. The outcome would have been different, in some measure at least, if farming had remained prosperous, and if the deluge of cheap wheat from the new countries had not arrived. In the event, the heavy-land farmer threw down most of his land to grass, and the scope for steam tillage was greatly reduced. Steam tackle became a useful supplement to the horse, but no substitute for it ; further important developments in mechanisation were to be delayed for a generation, until a lighter and handier power-unit had come into being.

In 1872 the triennial system of competitive trials was abandoned in favour of a nine-year rotation and, as it happened, nothing very important emerged from the trials of horse-drawn tillage implements in 1873, of row-crop implements and carts in 1874, or of hay-making machinery in 1875. In the last of these years some of the implement makers expressed dissatisfaction with the prize system, and there were criticisms in the press. The criticism was not maintained, but it does seem that, during a period when there was a pause in the stream of new ideas, the trials lacked point. They amounted to comparisons between well-known and established models, and the awards expressed little more than the individual preferences of the judges who happened to officiate.

About this time it began to be realised that important developments in agricultural engineering were in progress in the United States, and it was learnt that the Philadelphia Exhibition of 1876 would provide an admirable opportunity for the study of these. The Council accordingly sent John Coleman to attend the exhibition, and a full report by him appeared in the *Journal* for 1877. Here he described various labour-saving novelties—ploughs with new types of mould-boards and ploughs on which the ploughman rode ; a wagon-loader for hay harvest ; a maize-planter which could set the seed in a chess-board pattern to allow for two-way horse-hoeing ; a new type of thresher, and a variety of miscellaneous tools. The outstanding interest, however, attached to new types of corn-harvesting machines. There was a " header " which delivered the ears of the corn on to a travelling canvas, and carried them up by an elevator, so that they could be deposited in a wagon which travelled alongside. There was a reaper which delivered the cut corn on to a binding table, where three men gathered it and bound it into sheaves. But " the great feature of this department, and indeed of the whole Agricultural Exhibition, was the automatic binders, the realisation of a long-cherished notion on which mechanics have been engaged for years. At present the most successful of these inventions, of which four different kinds were shown in actual work, is probably far from perfect ; but just as the first Exhibition of 1851 was memorable for the introduction of a reaping machine, so will this be remembered as the first public occasion on which automatic binders were successfully worked."

The four machines which were tested—McCormick's, Wood's, Osborne's and McPherson's—all used wire as the binding material. Coleman confessed to a former prejudice against wire, but said that this had been largely removed by what he had seen. The Council acted on his report by publishing an offer of a Gold Medal, at the Liverpool Show of 1877 or at any future meeting, for an efficient sheaf-binding machine. Eight entries, of which three were American, were received, but the Howard, Burgess and Key, and Phillips machines were not ready to face the judges. The American entries were the wire-binding machines of McCormick, Wood and Osborne. Neal of London showed a string-binding mechanism which actually tied a knot, and King of Stroud another which twisted the ends of a string band. A field trial was arranged during harvest, but only the three American machines came forward. The judges came to the conclusion that they could not award the prize, but gave Wood a silver medal as a recognition of progress, and commended Osborne's machine.

The offer of a gold medal was still open, and at Bristol in 1878 there were seven entries—the wire machines of Wood and Osborne ; an improved and modified McCormick model made in England ; a wire-binding machine by Howard of Bedford ; and three machines using twine, one American and two English. This time the Judges had no hesitation in awarding the medal to the McCormick machine, and they commended Wood's ; but they added a rider to the effect that the use of wire was objectionable.

The next trial took place at Derby in 1881 ; Appleby's knotter had been invented in the interval and had been adopted by McCormick. There were originally twenty entries, but only eight came to the trials, and these included only four distinct types. The machines were judged on a score-card system, and McCormick's led with a score of 240 points and had the Gold Medal. Machines by Samuelson of Banbury (Oxfordshire) and by the Johnston Harvester Company of London were equal second, but these were both manufactured under the McCormick patents and differed from McCormick's own machine only in trifling details. The trial was thus a complete victory for the McCormick pattern. Hornsby and Howard both entered machines on this occasion but were out of the running ; indeed, they had come into the business rather late, and had still something to learn ; both, however, persevered, and in the next trials, at Shrewsbury in 1884, they came respectively first and second in a large field. Mr. Thomas Bell, who reported on this fourth series of trials, said :

> There may be room for greater simplicity and for further reduction in price, but I am free to state that the sheaf-binder can now cut and tie an average crop of grain in a manner superior to any other process of cutting and tying that I have yet witnessed. The machinery for elevating and binding is no doubt somewhat complicated ; but on this point there is not, in my opinion, anything approaching the difficulty which was present to the ordinary agricultural mind when manual and self-delivery reapers were introduced.

Mr. Bell's judgment was fully borne out by subsequent events for, from this time onwards, the use of binders rapidly spread. Unfortunately this invention was to prove anything but an unmixed blessing to the British farmer. It offered, for the first time, a means of exploiting the stores of fertility in the American prairies and its use, in the subsequent ten years, was to drive down the price of wheat to a level that ruined many heavy-land arable farmers in England.

From the 'seventies onwards dairying became increasingly important in Britain, and the production of milk for the liquid market became increasingly a matter for the dairy farmer as opposed to the city cowkeeper ; the year of the rinderpest, when so many of the cowkeepers were put out of business, marks the beginning of the change. It was natural, then, that the Society should have given increasing attention to dairying and especially to the more elaborate equipment that became necessary when milk was to be transported for some distance instead of being delivered, warm from the cow, from a cowshed in the consuming area.

The first trials of dairy utensils were held at Bristol in 1878, and there were classes for butter-churns and butter-workers, cheese vats, curd mills, etc., and also for milk churns suitable for long-distance transport, for milk coolers (a recent innovation but already being generally adopted) and

" for a method of keeping a large quantity of milk at a temperature of under 40° Fahrenheit for a period of not less than twelve hours " ; this seems to imply a farm refrigerating plant but, in fact, the winner used a supply of ready-made ice. The Judges had to regret that there was no entry in the class for milking machines. " He who successfully solves this difficulty will reap a rich reward. The want of such a machine is the one missing link in dairy management. Greater mechanical difficulties have been overcome, and we hope, before many years, to see the milking-machine difficulty solved."

One of the many wonders of the memorable if disastrous Show at Kilburn in 1879 was Laval's cream separator, exhibited for the first time in England. Another was a refrigerator wagon which won the special Mansion House Prize for a railway car " capable of conveying perishable goods a journey of 500 miles at a low temperature." The winning exhibit was nothing more than a well-insulated wagon with an ice-box, and with a fan for maintaining a circulation of cold dry air. It seems to have suggested to nobody the possibility of Argentine beef or New Zealand lamb and, indeed, another decade was to pass before this new cloud appeared on the home farmer's horizon.

In 1880, at Carlisle, the Society added a new feature to the Show in the form of a working dairy, in which the latest dairy utensils could be shown in actual use. It attracted a great deal of interest and was made a permanent institution. Trials of separators were advertised for the following year at Reading and there was an original entry of ten machines ; a patent case, however, was down for hearing in the Courts, and all but two of the machines were withdrawn ; the Laval separator had the award. A further and much more elaborate trial was held at Doncaster in 1891.

In the *Journal* for 1890 is a short note on the ' Murchland ' milking machine, recently invented by a sanitary engineer of that name who was in business in Kilmarnock. It was exhibited at Doncaster in 1891. This was by no means the first attempt to solve the problem, but it embodied the important new feature of a vacuum pipe, which was carried round the cowshed and had taps at convenient intervals for the attachment of the milking units. The vacuum was produced by a hand air-pump. Five years later, at the Darlington Show, a silver medal was awarded to the Thistle Company of Glasgow for a machine that introduced the further new feature of intermittent or pulsating suction. Other machines were shown in 1896, 1897 and 1900, but it was not until 1905 that the Society's Silver Medal was awarded ; the novelty of the winning machine consisted in the use of atmospheric motors suspended beneath the cow and worked in conjunction with double-walled cups. It was subjected to a two months' test at the Bedford County Institute, and performed so well that the Judges had no hesitation in making the award. Since those times there have been many refinements in construction which have, in sum-total, made for a large increase in efficiency ; yet it cannot be said that the milking machine has yet proved a complete or striking success. Machines have been installed and later have been discarded. The general view to-day is that the milking machine is an important labour-saving device in connexion with large herds but that, unlike the binder, it does not do its work better than that work can be done by competent hand labour. Probably a majority of farmers still take the view that the machine is to be preferred only where competent milkers are not obtainable in the numbers required.

We have seen, in the reaper and the binder, examples of the complete solution of particular problems that the mechanicians set themselves ; we have seen also, in the steam plough and the milking machine, examples of qualified or incomplete success after long-continued efforts. There are also to be found, in the pages of the old *Journals*, records of failure—failure in the sense that strenuous efforts to produce a new machine or devise a new process have been abandoned by the generation which made them, and the problem bequeathed to posterity. Perhaps the most interesting of these was the attempt, in the 'eighties, to devise improved methods of fodder conservation ; the whole idea was dropped, and has been revived only in our own time.

In the 'eighties, as fifty years later, it was realised that hay-making and root-growing, as methods of procuring winter stock food, were both open to objection. Hay-making depends on weather, losses are considerable even under favourable conditions and can be, at the worst, quite disastrous. Root-growing and root storage are costly in hand labour. If only grass, which is the cheapest source of food, could be conserved at a reasonable cost and without serious loss, the gain to agriculture would be very important.

In the 'eighties, as in our own time, two alternatives were tried : crop drying and ensilage. About 1880 a number of people claimed to have produced equipment for the artificial drying of hay and corn, and Mr. Martin J. Sutton offered, in connexion with the Reading Show of 1882, a prize of 100 guineas to any of these whose claims would be substantiated in an actual trial. The Corporation of Reading and Mr. Colebrook, a leading citizen, offered between them to sacrifice a hundred acres of hay crops in the cause of science, and an acreage of barley was also provided. There were eight entries, one machine operating by hot air on the unstacked crop, six by exhaust fans in the stack and one by forced ventilation of the stack by hot air.

In all the ninety-nine volumes of the *Journal* there can be few more entertaining official reports than that of Mr. W. C. Little in the issue for 1882. The weather was consistently wet ; the band of skilled hay-makers who had been brought from the Yorkshire Dales had to go home before the work was half done, and the unskilled crew of casuals who replaced them caused constant trouble ; the hay was largely spoilt and the trials upon corn were a disastrous failure.

The duty of the Judges would have been easier and more pleasant if they could, while declining to give the prize, have given a few words of encouragement to the exhibitors . . . ; but any such smooth words would misrepresent the opinions which they entertain. Bad and fickle as is this climate, they would far rather take the chances of weather than trust to any of the expedients brought under their notice.

The prospects for artificial drying being so unpromising, and a more hopeful alternative means to the same end being in sight, we hear little more of crop driers until 1925, when, at the Chester Show, a new type of plant gained a silver medal. The interest in ensilage is reflected in the *Journal* for 1884, when machinery for chopping and elevating green fodder was tested. No prize was awarded, though the cutter shown by Richmond and Chandler received high commendation. In the same volume is a long article by H. M. Jenkins, the Secretary, on *The Practice of Ensilage at Home and Abroad*, as well as a paper by Dr. Voelcker on the chemistry of the process.

From 1883 onwards, except during the years of the War and the immediate post-War period, the policy of the Society has been to encourage inventions of all kinds, and from time to time, as occasion seemed to demand, to carry out systematic trials of particular classes of implements. Inventors have been invited to enter new implements for the Society's silver medal, and medals have been awarded in cases where, in the Judges' opinion, the implement embodied a new principle and was of sufficient importance to farming. In recent years all entries for the medal have been subjected by the Society's engineer to exhaustive tests. A very large number of systematic trials have been carried out and, especially in the period between 1890 and 1914, two or more types of machines were sometimes included in a single year's programme. The mere enumeration of the awards and results would occupy many pages, and any attempt to trace out the various lines of development of modern machinery would involve the writing of a book. None, however, will deny that the most important addition to the equipment of the farm in the present century has been the tractor, and we may conclude this chapter with a short account of its evolution.

Stationary internal-combustion engines began to appear at the Shows in the 'eighties and the Judges at the Nottingham meeting, in 1888, were sufficiently impressed to award a medal to a paraffin engine exhibited by Priestman Bros. of Hull.   The special merit of this was its ability to run on ordinary kerosene as opposed to the low-flash-point oil which had been the fuel of earlier models, and which was regarded by fire-insurance companies as dangerous.   Another engine of the same type, but portable, had a medal in the following year.   From this time onwards the new power unit began to attract the notice of farmers by reason of its quick starting and general convenience as compared with the steam engine.   In 1894, at Cambridge, the Society ran a trial which attracted an entry of 26 engines, 17 fixed and the remainder portable.   The trial showed that there was already a wide choice of efficient engines, though Hornsby's and Crossley's were superior, in all-round merit, to the others.

In 1897 at Manchester, and again in the following year at Birmingham, prizes were offered for self-moving road vehicles driven either by oil or by steam.   In the class for light-load vehicles (up to 2 tons) were three motor-vans, two of which used petrol and the third paraffin.   Of these, however, only one, entered by Daimler, completed the forty-seven-mile road course set by the judges.   Its average speed was 7·8 miles per hour.

The first farm tractor that was considered worthy of mention by the Society's implement judges appeared at the Park Royal Show of 1903. This was the " Ivel Agricultural Motor," a three-wheeled model of about fourteen horse-power, built by Mr. Daniel Albone of the Ivel Cycle Works, Biggleswade.   Its forward speed could be varied by changing a sprocket pinion, and there was a reverse gear.   No award was made at the Show, but in the autumn the judges went to Bedfordshire to see two of the machines at work, one hauling a lorry on the road and the other pulling a three-furrow plough.   They were satisfied that the tractor could do not only these and other similar tasks but that it might be used as a source of power for driving threshers and other stationary machinery.   Their main objection was to the use of petrol, at that time considered to be a highly dangerous material to be stored or used on the farm.   An improved model of the Ivel, with two forward speeds, a radiator and automatic lubrication, was exhibited at the Show of 1904, and this was awarded a silver medal.   In 1905 the Ivel Company, which had been formed in the interval, showed a carburettor which allowed either petrol or paraffin to be used.

XIXa.—The Winning Vehicle in the Trials of 1898
This van completed the 47-mile course in 6 hours

XIXb.—The Ivel Agricultural Motor
awarded silver medal, 1904

A. CASE 10–18 H.P. FIRST PRIZE, CLASS I

B. CLETRAC. SECOND PRIZE, CLASS I

C. MANN STEAM TRACTOR. FIRST PRIZE, CLASS IV

D. CRAWLEY MOTOR PLOUGH. FIRST PRIZE, CLASS VII

XX.—TRACTORS AT THE LINCOLN TRIALS, 1920

To the same show Messrs. Saunderson of Bedford brought their 'Universal' Motor, a four-wheeler with independent drives to the front and rear wheels. This could be used to draw ordinary implements, could be attached to a special plough or could be turned into a four-ton lorry by fitting it with a special body. The judges noted certain defects, and deferred judgment until the inventor had had time to remedy these. An improved model, with a carburettor capable of dealing with paraffin, was exhibited at Derby in 1905, and was awarded the Society's medal.

Early in 1909 the Society decided to offer a gold medal for the best Agricultural Motor to be presented for trial in 1910, and the trials took place at Bygrave, in Hertfordshire, in August of that year. The specification was for " any form of motor, using steam, oil, petrol or electricity as its motive power which

(a) shall be capable of hauling direct in work a plough, cultivator, harvester or other agricultural implement ;

(b) shall be capable of driving such agricultural machines as a threshing machine, chaff cutter, grist mill, etc. ;

(c) shall be capable of hauling a load along a road or on the land."

There were eleven entries, but of these only seven came forward ; they included four steam tractors, the Ivel motor and two Saunderson models of 45–50 and 25–30 horse-power respectively. The trials included ploughing, harvesting, traction and brake tests, and a full report was published in the *Journal* for 1910. All the tractors accomplished the tasks set, but the judges " were unable to recognise in any of them the agricultural motor which is hoped for as the ideal general-purpose tractor and engine for farm purposes." They were, however, of the opinion that the trials would prove useful in leading to the development of such a motor, one which, while not being heavy, would enable the farmer to dispense with the hiring of a traction engine, and would be capable of doing a substantial amount of field work.

During the next few years there were many experiments. In 1912 the Daimler Company showed an enormous tractor of 105 horse-power, said to be capable of drawing a twenty-one furrow plough. In 1914 came a Motor Hoe from the Ivel Company, and a light, low-powered Saunderson tractor. By 1915 makers seem to have abandoned, for the time being, the idea of a general-purpose tractor, and the most interesting exhibits were three motor ploughs, all designed in such a way that the driver sat behind the plough and had controls reaching to the engine in front. Two of these motor ploughs were carried on, and driven by, a single pair of wheels, while the third was provided with caterpillar tracks. In the *Journal* for that year, also, is a review by Mr. Arthur Amos of the whole problem of tractor ploughing ; in the same year we note a decision by the Council to hold full-dress trials as soon as possible after the end of the War. These trials took place at Aisthorpe and Scampton, near Lincoln, in the autumn of 1920. There had been, in the meantime, important developments in connexion with the war-time food-production campaign, and numbers of American tractors had been imported by the Food Production Department of the Board of Agriculture. But steam power was not yet out of the running, and it was also still uncertain whether the most satisfactory plan of applying mechanical power in tillage was by direct traction, by cable haulage or by means of an engine and implement combined as a single unit. Seven classes were made to provide for these various possibilities ; three were for internal-combustion tractors capable of drawing two-, three- and four-furrow ploughs respectively ; one was for steam tractors capable of taking four-furrow ploughs ; one for double-engine cable sets using oil engines and one for the corresponding type of steam plant ; finally, there was a class for self-propelling ploughs. A gold and a bronze medal, with cash prizes, were offered in each class. Forty-six machines were entered, thirty-eight came forward and thirty-six completed their allotted tasks.

It is impossible here to summarise the long and fully illustrated report of the judges, which appeared in the *Journal* for 1920. The broad conclusion was that the farmer had a wide choice of excellent tractors, combining economy in fuel consumption with handiness and reliability, and covering a wide range of size and power. The trials cost the Society nearly £5,000, but they materially helped to establish the tractor in general use, and provided a mass of accurate data on which the individual farmer could base his choice of a particular model.

The next tractor trials took place in 1930 and were carried out with the co-operation of the Institute for Research in Agricultural Engineering of Oxford University. Entries were invited from all countries and of the 31 which appeared 12 were American, 8 British, 4 French, 3 German, 1 Irish, 1 Hungarian and 2 Swedish. These included paraffin, petrol, diesel and semi-diesel types. Entries of cable sets and of self-propelled cultivating machines were invited, but none actually appeared. The tests, consisting of belt tests, draw-bar tests and field tests (ploughing and cultivating), were carried out by the Institute and occupied a period of seven weeks, after which there was a public demonstration. The printed report of the tests was available to visitors, and some 1,700 copies were sold on the ground.

In 1937 the present scheme came into force. Its object is to provide a permanent means for the testing of all tractors intended for sale in Britain. The tests are carried out on practical lines so as to provide the kind of information that is of direct value to the farmer. The scheme recognises, on the one hand, that draw-bar work in cultivation is the most important function of the tractor and, on the other, that the information most urgently required by farmers, in regard to any particular machine, is the rate of working and the fuel consumption in ordinary farming operations. At the same time the scheme provides for measurements of draw-bar pull and the width and depth of working, so that accurate comparisons can be made with the performances of the machines in more formal engineering tests and in work on other types of soil. Ten tractors were tested in 1937 and fifteen others in 1938. The results are published annually in the *Journal*.

The progress of agriculture is largely an affair of obtaining from the soil a larger return from a given amount of human effort. In the past century progress has been made on a wide front, so that the farmer of to-day has much knowledge and many resources that were lacking in 1839. It is difficult to assess the relative importance of discovery and invention in the different branches of science—to compare, for instance, the benefits derived from better fertilizers, better plants, better animals and better machines. But the development of farm machinery has been, at least, one of the main lines of advance, and in this the Royal Agricultural Society has played an important part. It has doubtless made mistakes. Its judges, at times, have erred—both by raising false hopes and by refusing to recognise important possibilities. The Society's policy has at times been criticised both by the manufacturer and inventor on the one hand and by the farmer on the other. But, upon the whole, it can hardly be denied that the Society has well performed the task that it set itself—" To encourage men of science to the improvement of agricultural implements."

CHAPTER VIII

VETERINARY SCIENCE AND ANIMAL DISEASE

THE eighth of the Society's objects, as originally set down, was " To take measures for improving the veterinary art as applied to cattle, sheep and pigs." The first report of the Committee of Management shows that its members were already alive to an unsatisfactory situation—to the heavy losses suffered by farmers through live-stock disease, and to the poverty of knowledge on the subject.

A Veterinary College had indeed long been established near London, but the veterinary surgeon of those days was in fact, as he was often called, a " horse doctor " and the College had, accordingly, given little attention to any other class of stock. One of the first actions of the Committee was to approach the Governors of the College with an offer of funds for the provision of instruction in the pathology of cattle and sheep. Although the Principal of the College, Professor Coleman, seems to have been rather unenthusiastic about the proposal, the offer was favourably received by the Governors, the Society appointed delegates to carry out the arrangements and an annual grant of £200 was made. Thus began an association which, although it was not to remain unbroken, was to produce valuable results.

By a strange coincidence there appeared in England, during the time that this arrangement was under discussion, a new and serious disease of cattle, sheep and pigs. The Society's Committee immediately arranged for an enquiry, and obtained the assistance of Professor Sewell of the College. The first volume of the *Journal* contains his preliminary report : the symptoms are described as a feverish condition with severe local inflammation and the formation of vesicles on the lips and tongue and between the claws of the feet ; the inflammation might start either in the mouth or on the feet, but always spread from one to the other. Foot-and-mouth disease had arrived.

Professor Sewell gave directions about treatment and his suggestions, amounting to little more than careful nursing, seem to have been sound enough. The Committee circularised members of the Society asking for detailed accounts of outbreaks, and of any methods of prevention or treatment that had met with success ; but, as may be supposed, nothing of particular value emerged from the replies. It was agreed that the malady generally spread by contagion, but many of the outbreaks could not be so explained. Breeders of store cattle were not greatly perturbed about the disease, for the death-rate was small and the loss of condition that resulted from an attack was quickly made good. The London dairymen were harder hit, for the disease was very prevalent in the city cowsheds and it was reckoned that a freshly calved cow, if she became affected, might lose about a third of her value. It is clear, however, that the British farmer's first experience of the disease must have been of one of its milder forms.

Two years later, in 1841, another unwelcome immigrant arrived in the form of Contagious Pleuro-pneumonia of cattle. This disease was generally supposed to have come from Holland, and there is no doubt that it had been rampant there for some time ; but it is also a fact that it was recognised in Ireland before the first English case was reported. It may be, indeed, that this was not the first outbreak that England had known, for the descriptions of the murrains and plagues that affected the cattle of this country in earlier times are so vague that the diseases cannot be identified with certainty. In any case England had been free for a long period up till 1841, but from that year the malady continued to spread, and every outbreak resulted in a high proportion of deaths. In 1847 the Council made Pleuro-pneumonia the subject of one of its prize essays. The £50 premium was awarded to George Waters, a veterinary surgeon in Cambridge, and his essay appeared in Volume IX of the *Journal*. Waters gave a very full account of the symptoms of the disease. He also pointed out that the droving of imported cattle through the country, and their exposure at markets and fairs, must lead to its spread ; that cattle bought in public markets should be isolated on reaching the purchaser's farm ; that infection persisted for some time in sheds or yards that had been occupied by diseased cattle, and that, therefore, thorough disinfection of such premises was highly important.

By 1851 pleuro-pneumonia had " spread far and wide in this country, and destroyed great numbers of our cattle." In that year the Council heard of some promising experiments, carried out in Belgium, on the preventive inoculation (vaccination) of cattle, and they sent over Professor Simonds (who, in the interval, had been appointed as the Society's Veterinary Inspector) to investigate and report. He found that the method consisted in making a deep incision into the tail of the animal that was to be immunised, and inoculating the wound with fresh lymph from the lungs of an infected animal. The treatment, when it succeeded, caused a purely local infection which led to what we should now call active immunity, and so protected the animal against the very serious and often fatal lung infection. Some deaths generally followed vaccination, and numbers of animals lost their tails, but these losses were small by comparison with the wholesale deaths that occurred when the disease was allowed to run through a herd. Simonds reported that the method was promising but that more experiments were necessary. The Council accordingly arranged that he should have facilities for trials in this country and these were carried out. As a result of his experiments Simonds reported against vaccination ; he may have been right so far as England was concerned, but the treatment came to be widely used in other countries.

Still another fatal disease of live stock—sheep-pox—was introduced, probably from Holland, about 1847 ; by the middle of the century it was causing considerable losses. Mr. Stanley Carr, the English manager of a large estate in Schleswig, wrote to the editor of the *Journal* giving his own and his neighbours' experience of the disease, and especially of the method of inoculation used against it ; shortly afterwards Simonds published a fuller account of the process. But it was now becoming clear to British farmers that the importation of live animals from the Continent must be a continual source of danger to the health of the country's live stock ; later on, the Society was to fight a long battle on this issue, and eventually to win ; but the battle was not yet joined.

To revert to the Council's desire for the broadening of Veterinary Education, we find that, by 1847, there was dissatisfaction with the response that had been made by the Veterinary College. It may be, as the Council seems to have thought, that the College authorities were apathetic. On the other hand, it must have been difficult for a lecturer in cattle pathology, with the best will in the world, to find much information that was worth imparting to his students. Veterinary medicine was still an art rather than a science ; the teaching of the College must have been based mainly on the accumulated experience of practitioners ; and qualified veterinarians were very rarely consulted in cases of cattle disease and were scarcely ever called in to sheep or pigs. However this may have been, the Council gave notice that their grant to the College would be discontinued in 1848 and for the next two years it tried other means to the end that it had in view.

At this period the policy of offering substantial prizes for essays, on subjects chosen by the Council, was producing much really valuable material for the *Journal,* and from now on Veterinary subjects began to appear on the lists. One of these has been mentioned above. The subjects chosen for 1849 included " Diseases of Cattle and Sheep occasioned by Mismanagement " and " Abortion in Cows " ; and those for 1850 " Diseases following Parturition in Cattle and Sheep." These essays illustrate very well the state of veterinary knowledge at the time, showing, on the one hand, how much valuable empirical knowledge had been accumulated and, on the other, the frequent hopelessness of finding remedies for a disease so long as its essential nature was not understood. The writer on the first-mentioned subject records many observations on such points as the relation between liver rot and wet pastures ; between the consumption of frosted pasturage and outbreaks of braxy in sheep ; and between various factors in management and the incidence of husk in calves. The author of the essay on Abortion distinguishes clearly between contagious and accidental abortion, points out the risk of introducing infected cattle into healthy herds and urges the importance of isolating cows that have aborted or show signs of approaching abortion. On the other hand, the long account of Milk Fever gets us nowhere, and the various complicated treatments recommended, according to the symptoms of the particular case, seem to the modern reader about as likely to kill as to cure.

At this time also the appointment of Professor Simonds as the Society's Veterinary Inspector was made. He was to visit the farms of members where there were outbreaks of epidemic disease and to give advice, the Society paying his fees and the member reimbursing his travelling expenses. He was also to be on duty at the Society's Shows in order to examine the exhibits for hereditary defects and contagious disease and to check, by their dentition, the ages of the animals shown. Professor Simonds made a very detailed study of dentition from this point of view and, as a result of his experience, became the leading authority. The long paper that he wrote for the *Journal* is probably the first authoritative account of the subject.

In 1852 the Council was able to report that it had composed its quarrel with the Veterinary College, and that the Society's grant would be renewed upon amended conditions. Members of the Society were to have the privilege of sending cattle, sheep and pigs to the College for observation, treatment and *post-mortem* examination ; the College undertook to investigate particular diseases, as the Council might direct ; Professor Simonds, as the holder of the lectureship on Cattle Pathology which the Society endowed, was to give an extended course of lectures and was also, on request, to deliver lectures to the Society ; lastly, the College was to furnish the Society with reports on the cases which reached it from members of the Society.

In 1855 the distinguished veterinarian Finlay Dun appeared as a contributor to the *Journal*, and the ' Royal ' could congratulate itself that the Profession was beginning to give serious attention to those aspects of veterinary science that were of interest to the general farmer as well as to those that concerned the horsekeeper.

In 1856 another cloud appeared on the horizon. Rumours began to reach this country of a very fatal disease of cattle on the Continent. Early in 1857 the attention of Parliament was called to the subject, and the Foreign Office undertook to collect reports from British Consuls in European countries. These reports confirmed that epidemic disease in cattle was rampant and that, in some places at least, the malady was the dreaded steppe murrain or rinderpest. The exact position, however, was not clear, and it was considered desirable to send a qualified veterinarian to investigate. The three National Agricultural Societies therefore combined to send Professor Simonds on a continental tour. His long and detailed report appeared in the *Journal* for 1857 and was very reassuring. He had found that all the countries of Northern and Western Europe, from which cattle were exported to England, were perfectly free from rinderpest ; that the only really serious epidemic disease prevailing in these countries was pleuro-pneumonia, which the British farmer already knew ; that many of the consuls had obviously confused rinderpest with pleuro-pneumonia or anthrax ; that rinderpest was a disease of the steppes which had, indeed, occasionally crossed the Russian frontiers into Hungary, Poland and other neighbouring countries, but that these countries were alive to the danger, and were taking adequate precautions ; and that, speaking generally, rinderpest had not been known in Central or Western Europe for a period of forty-one years. Steppe murrain was, indeed, a disease greatly to be dreaded, for it was highly infectious, and the death-rate was often ninety per cent. of the herd attacked ; but " No fear need be entertained that this destructive pest will ever reach our shores." Eight years later Professor Simonds himself was to diagnose the first case of rinderpest in England.

English farming had made a quick recovery after its temporary setback in the early 'fifties, and was now in a very flourishing condition. The Society shared in the prosperity of the industry, and the Shows of the period were mostly very successful from every point of view. The Veterinary Committee of the Council seems to have had little business of importance. The Governors of the College complained that they were hampered by the failure of members of the Society to take advantage of their privileges ; its students were now showing increasing interest in the pathology of cattle and sheep, but there was a great shortage of clinical material for teaching, and it was to be wished that members of the Society would send more cases to the College for treatment and report. In response, a statement of members' veterinary privileges was inserted in the *Journal*, but the complaint was repeated in the same terms a few years later.

The task of examining the show stock for hereditary unsoundness, and of verifying the ages of the animals, had now become too heavy for one man, and Professor Spooner, of the Royal College, was appointed Joint Inspector with Professor Simonds. The latter was still very active. He carried out a very thorough investigation of liver rot in sheep and, although he failed to trace out the very complicated life-history of the parasite, his long report in the *Journal* was a valuable contribution to knowledge. Among his other contributions was a useful one, in the *Journal* for 1865, on the external parasites of animals, including an account of the most up-to-date methods of sheep dipping.

In 1865 the Show was held at Plymouth and, despite the fact that the country was in the midst of a general election, drew a very large attendance, and " no dread of coming calamity disturbed the harmony of the occasion." The events of the next few months are described by Howard Reed in the following year's *Journal*. His account of the time is so vividly descriptive that it may be quoted at length :

Returning from this meeting the exhibitor could scarcely have crossed the threshold of his home before the Lords in Council issued their first Order, which announced the recent appearance of " a contagious or infectious disorder of uncertain nature, prevailing within the metropolis and in the neighbourhood," and advised measures to be taken immediately " to prevent such disorder from spreading." Within six months from that date anxiety has deepened into despair. Thirty-six out of forty English counties are infected ; in Scotland, sixteen out of twenty-seven ; in Wales, two out of twelve. The depressing totals of the bills of mortality have doubled from month to month. . . .

The rinderpest was detected in its first appearance towards the end of June in two metropolitan cowhouses by Professor Simonds, and reported to the Home Office on the 10th of July. Having formed a previous acquaintance with this murrain, in its haunts and special breeding grounds, Mr. Simonds was well aware of the impending danger, and accordingly sounded an alarm of which, at first, but faint echoes were heard through the country. This timely proceeding gained for him the title of " Alarmist " with those who were ignorant of events from 1745 to 1757,[1] and unhappily they were many ; but the event has proved that his fears

---

[1] The time of the last preceding outbreak of rinderpest.

were founded on correct knowledge of the subtle and deadly nature of this fever. Before the close of July Professor Gamgee, in addressing a metropolitan meeting of the troubled cowkeepers, announced that 2,000 cows had already perished, though the disease was but a month old. Throughout August the newspapers were pretty fully employed in chronicling its progress through the provinces, for it appeared with a simultaneousness that quite confirmed the views adopted by some superficial advocates of "spontaneous generation." . . . Rayed out from the Metropolitan Market it multiplied its centres of operation, and wrought with a rapidity known only to itself. The Veterinary schools set to work to observe and advise. They made *post-mortem* examinations, did their best to establish the identity of the murrain, wherever it appeared, with the rinderpest of Russia, and when they had done so agreed that "to slaughter and to stamp out" was the only prescription of any avail. Town and country meetings were held in the various districts where the disease appeared, to concert measures of defence. From these sprang Mutual Insurance Associations and Medical Commissions, and other organised operations, the first provincial examples of which were those of Norwich and Aylesbury. Under the pressure of public opinion, medical testimony, and the exigencies of the case the Privy Council issued no less than five distinct orders during the month, the last of which empowered justices to appoint inspectors authorised "to seize and slaughter, or cause to be slaughtered, any animal labouring under such diseases."

But, in spite of this Order, the pest raged with redoubled violence throughout September. . . . Unfortunately, too, the order in certain cases increased the mischief it was intended to subdue ; for the inspectors, being men often singularly unfit for the office, yet unwearied in their efforts to perform the despotic duties attached to it, carried the disease wherever they went, and left it where they did not find it. The fact that men were to submit to the slaughter of their cattle without receiving compensation did much to spread the infection, for where the disease appeared the owner felt no compunction about clearing off the greater portion of his stock before hoisting the black flag. . . .

During the same period the Privy Council consolidated the previous orders, prohibited the importation of hides and skins, sheep, lambs and all cattle into Ireland ; forbade the entrance of cattle into the Metropolitan Market, except for the purposes of being slaughtered ; empowered magistrates in Petty Sessions to stop fairs and markets ; and published papers containing suggestions on the part of Professor Simonds, and a memorandum on the principles and practice of disinfection, by Dr. Thudiehum.

At the close of September the Privy Council, finding its efforts to stay the plague unavailing, solicited the Queen to issue a Royal Commission to investigate the origin and nature of the disease, and to frame regulations to check its progress.

Throughout October, during the laborious daily sittings of the Commission, the plague continued to extend. . . . The doctors and nostrum-mongers were of course busy. Multitudes had a specific, but no one held a cure. The practitioners being ill acquainted with the nature of the disease . . . treated it from various points of view. The results, however, were dismally uniform. . . .

The Report of the Commission, when it appeared in November, brought no relief. It rather increased the alarm, for the evidence on which it was founded confirmed the gloomiest forebodings . . . and pointed out a remedy in slaughter and the absolute stoppage of cattle traffic which the Government . . . did not feel warranted to enforce before the meeting of Parliament. They preferred to issue another Order, revoking powers previously given to inspectors and stimulating local action by conferring more extensive powers on local authorities. Unhappily this step induced an amount of confusion of which the plague took the most ample advantage ; for throughout the months of November, December and January, our loss doubled every four weeks, and at the close of February it amounted to nearly 12,000 a week.

The situation in February 1866 thus looked about as bad as it could be.

Early in the course of the crisis the Council of the Society had formed themselves into a Standing Committee and had kept constantly in touch with the situation. As a result of their study of the problem they drafted a series of resolutions and, on February 12, they had an interview with Earl Russell, the President of the Council, and put the resolutions before him. Since these formed the basis of the Cattle Diseases Prevention Act, which was passed in the same month, and since they set out the policy that the Society was to continue to urge in relation to epidemic diseases generally, they may usefully be given in full.

Resolved—That the Cattle Plague has increased and is increasing to an alarming extent. That the measures hitherto adopted have been wholly ineffectual to prevent its progress.

That no method of dealing with the Cattle Plague, at the present time, will be of any avail unless it provides for :

1st. The immediate slaughter and burial, at least 6 feet deep, of all cattle suffering from the disease ; making compensation to the owners in such mode and to such extent as shall be considered advisable.

2nd. The rigid surveillance of all infected farms, and the immediate slaughter of all animals which, from time to time, shall show the slightest symptoms of disease.

3rd. The thorough disinfection of all infected farms, and a prohibition to remove therefrom all manure, litter, hay or straw for a period to be fixed ; and then only subject to certain specific regulations.

4th. That the Government be requested to bring before Parliament a Bill to direct and empower Justices in Quarter or Special Sessions to assemble immediately to carry out the above Resolutions ; and in such Bill make provision for charging the necessary expenses on the

County Rate, and also for assimilating the action of Counties and Boroughs.

5th. That simultaneously with the destruction of diseased cattle, the transit of all animals, whether by road or rail, be entirely prohibited, with such exceptions only as shall be absolutely necessary.

6th. That during the existence of the Cattle Plague, all imported cattle, sheep, or swine shall be slaughtered forthwith at the port where they are landed ; and their hides, skins and offal disinfected there.

With the bringing into operation of the new Act, the situation rapidly improved. In the week ended February 20, 17,875 cattle were returned as infected. The number for the fourth week of April was 4,442 ; for the last week of June 338 and for the last week of December only 8. The position was so reassuring that the Society proceeded with its plans for the Bury St. Edmunds Show, postponed from the previous summer. In the end it was thought wise to abandon the cattle section, but before the show date arrived the disease was at an end. It only remained for the Society's Council to urge that all the machinery which had been set up under the Act should be kept in being.

Despite the extra vigilance of the Government after this severe lesson, the position remained a precarious one for many years. The general disorganisation caused by the Franco-Prussian War gave the rinderpest a chance to spread widely on the Continent, and in 1872 it was rampant in Germany, Holland, Belgium and parts of France. In July 1872 lots of affected cattle were landed at Deptford, Hartlepool and Leith and, in one case at least, only a fortunate accident (that some of the cattle had foot-and-mouth disease as well) prevented their being admitted. On the 16th of January 1877 a further lot of cattle, shipped at Hamburg and landed at Deptford, were found to be affected. These were immediately slaughtered, with the usual precautions in the way of sterilizing the carcases and the lairage. Nevertheless, on the 29th of the same month there was an outbreak at Limehouse. The holding of markets in London was prohibited and the other usual precautions were taken, but within a week more herds were infected. The ' Royal ' Council, assembled for a special meeting on February 20, were told of a separate outbreak at Hull. This time, indeed, the disease was not allowed to get out of hand, but there were in all forty-seven separate outbreaks, and the situation for two or three months was one of intense anxiety.

Meantime, scattered outbreaks of pleuro-pneumonia had come to be regarded as normal, while foot-and-mouth disease flared up periodically —notably in 1871 and 1875. Some figures for the West Riding of Yorkshire, for the years 1870 and 1871, may be quoted to indicate the state of affairs. In this period 359 cattle were returned as infected with pleuro-pneumonia and of these 229 died or were slaughtered ; over 45,000 animals suffered from foot-and-mouth disease and of these 356 died or were slaughtered. Valuable pedigree herds, naturally, suffered with others ; for instance, the famous Booth herd of Shorthorns, at Warlaby, was infected with foot-and-mouth disease in 1872–73 and only three survived out of the whole season's crop of calves.

It is impossible, in the space allotted to this chapter, to describe in detail the efforts of the Society, between 1870 and 1895, to persuade the Governments of the day to assume more powers, and to adopt stricter measures, for the control of epidemic disease. The method employed was simply to bombard the Government Department concerned with petitions and, with unwearying perseverance, to pester ministers with deputations. In its earlier efforts the Society often acted alone, but as time went on it aimed more and more at educating and uniting agricultural opinion throughout the country, and at securing the co-operation of the other national and local agricultural societies. Up till 1889 the Government Department concerned was the agricultural committee of the Privy Council and thereafter the Board of Agriculture. It is not implied that either the one or the other ever turned a deaf ear to the representations that were made, but there were often difficulties to be overcome ; there were the vested interests of the butchers who dealt with imported cattle, and of the city dairymen who wanted Dutch cows ; and there was always the suspicion, on the part of the general public, that the farmer was fighting for protection against the competition of overseas producers rather than for protection against disease. Moreover, the farmers themselves were not always unanimous, and it was sometimes suggested that the ' Royal ' acted in the interests of pedigree breeders rather than in those of the industry as a whole—it being, of course, obvious that pedigree breeders have more to lose from disease than the general run of farmers.

It is perhaps sufficient answer to all the doubts and objections that were expressed to say that events have proved the wisdom of the policy which the ' Royal ' so consistently and persistently urged ; the inconvenience suffered through the operation of the Diseases of Animals Acts, and the costs of their administration, are far more than outweighed by the benefits which this country enjoys through its relative freedom from live-stock epidemics.

One chief aim of the Society's policy was to control imports of live animals from countries where serious epidemic diseases existed. The first step was taken as early as 1867 when an Act of Parliament gave powers to the Privy Council to define ports, and parts of ports, for the landing of foreign animals. This made possible the institution of a proper system of inspection which, however, as we have already seen (p. 112), did not always achieve what was hoped. A further Act of 1869 gave power to the Privy Council to prohibit imports, or to slaughter or quarantine imported animals, as they might think fit. Another Act of 1886 provided that the Privy Council should " prohibit the landing of animals from any country, or any part thereof, whenever they were not satisfied that the circumstances were such as to afford reasonable security against the introduction of foot-and-mouth disease."

Precaution could go only one step further, and the Society continued to pursue its object. In February 1893 Mr. Richard Stratton moved in the Council " that in the interests of the producers and consumers of meat in the United Kingdom it is essential, as a safeguard against the introduction of foreign contagious diseases with animals, that all cattle, sheep and swine imported into the United Kingdom from foreign countries, which are not for the time being ' prohibited countries,' be slaughtered at the port of debarkation except in special cases, when they may be admitted under such conditions as the Board of Agriculture may from time to time consider necessary."

This proposal was not embodied in the (otherwise excellent) Act of 1894 and therefore, two years later, when a change of Government offered a new opportunity, the Council returned to the attack. They got together a large deputation, representing the ' Royal ' and sixty-two other agricultural associations, which put the case before the President of the Board. This had the desired effect, and the proposal was embodied in the Act of 1896.

The second main object of the Society was to secure compulsory slaughter, with adequate compensation, whenever it appeared that such a policy seemed likely to lead to eradication of a disease. The first few weeks of rinderpest had shown that slaughter might offer the only hope of eradication, and that concealment of disease was inevitable so long as compensation was withheld. Compensation was in fact made compulsory, but the cost at first fell upon the local rates. This was unsatisfactory because not all local authorities were assiduous in their duties ; and it was inequitable because the object was national, and a disproportionately heavy cost might at any time fall upon a single town or county. The " Pleuro-pneumonia Act " of 1890 transferred the powers and duties of local authorities, in connexion with slaughter and compensation, to the Board of Agriculture.

Within the next few years it seemed that the situation with regard to pleuro-pneumonia was well in hand, that little further expenditure need be anticipated in dealing with the occasional scattered outbreaks that were likely to occur, and that the country could look forward to its complete eradication within a few years. These hopes were in fact to be realised. By the Act of 1892 the Board of Agriculture was empowered to use the money which had been voted for compensation in cases of pleuro-pneumonia to meet the costs of dealing with foot-and-mouth disease. Since that time, and in spite of increasing difficulties due to the growth of traffic with the continent, the system of control has prevented the disease from becoming permanently re-established. More than once the situation has seemed to be getting out of hand, and there has been agitation for the abandonment of the slaughter policy. In every such case the Society has supported the authorities in their firm action, and this, with a steadily improving organisation, has won the day. A French writer, discussing the widespread outbreak of foot-and-mouth in 1937–38, and the British campaign of eradication, pays the following tribute to our British organisation :

The tactics adopted by the Ministry of Agriculture have been subjected to a severe test. They have triumphed over all obstacles, and the courageous campaign, pursued with tireless tenacity, reflects the greatest possible credit upon the British Veterinary service. The total sums paid in compensation to owners, for the slaughter of stock in the 267 outbreaks, have amounted to £344,721, an insignificant sum if one compares it with the losses inflicted upon the vast districts on the Continent which have been invaded by the disease.

After 1894, there was a marked decline in the number of petitions and deputations to the authorities concerned with disease control. In 1919 there was a deputation on Sheep Scab, and this was followed by the new Sheep Scab Orders of 1920 and 1923. The position, however, remained unsatisfactory, and in 1925 we find the Council calling the Ministry's attention to the failure of their control measures and urging that sheep-scab be given more serious and prompt attention. Still there was little progress, and in 1929 the Society joined with the National Sheep Breeders' Association in sending a deputation to the Minister. As an outcome of the discussion the Society instituted a general propaganda campaign, circulating county officials, agricultural societies, Farmers' Union branches and other bodies, and making use of the agricultural press to reach the general body of farmers. The appeal was one for concerted action among all concerned and this did succeed in arousing a general state of greater vigilance and activity. In 1926 there was a resolution on the necessity to prohibit the importation of fresh animal carcases, a serious outbreak of foot-and-mouth having been traced to Belgian pig carcases imported for curing in Britain. The desired step again followed. But in the main, the Society's rôle in recent years has been that of a stout supporter of the firm measures which the Ministry of Agriculture has adopted.

We must now return to the 'seventies and follow the Society's efforts to promote veterinary education and research. In 1871 there had been established in London, under the *ægis* of London University, a body called the Brown Institution, founded for the purpose of investigating the diseases of domesticated animals. In 1875 there was another break between the Society and the Royal Veterinary College, on the old question—the alleged lack of interest, on the part of the College, in the problems of cattle disease. In this year the Society withdrew its grant from the College and gave to the Brown Institution a sum of £500 to be devoted, over a period of two years, to the investigation of pleuro-pneumonia and foot-and-mouth. The grant was continued, at the same rate, for five years in all, and the scope of the research was extended to include anthrax.

At this time the work of Pasteur was exciting great interest, for it was obvious that the discovery of the organism of anthrax had laid the foundations of a new science. Moreover, Pasteur had shown that a weakened strain of the bacillus could be produced by artificial culture, and that this weakened strain could be successfully used for vaccination. A report on the use of this vaccine, by the Principal of the Brown Institution, appeared in the *Journal* for 1881. Anthrax, or " splenic apoplexy," had been, up till this time, a frequent cause of loss to British stock-farmers, and vaccination proved to be a valuable preventive. In the end, however, the disease was kept in control by other means.

The year 1879 is one of evil memory, and among the various catastrophes which it brought to British agriculture was an outbreak, on an almost unprecedented scale, of liver rot. Estimates of the loss exceeded three million sheep. In the following year the Council made a grant of £100 to an Oxford Zoologist, Mr. A. P. Thomas, to finance an investigation of the disease. Thomas was successful in tracing the extraordinarily complicated life-history of the fluke, and showed quite clearly the connexion—through the water snail—between outbreaks of liver rot and the wetness of the land or the season. His reports appear in the *Journal* for 1881 and 1882.

The quarrel with the Veterinary College was composed in 1879, and in 1890 the grant was raised (to £500 per annum) in order to establish a chair of Comparative Pathology and Bacteriology. The grant remained at this level until 1903, when the disaster of Park Royal necessitated its reduction to the old figure of £200 ; in 1911, with the return of more prosperous times, it was raised to £400 and at this level, except for a temporary reduction in 1921–23, it has since remained. The Society also contributed £1,000 to the fund for the rebuilding of the College. The ' Royal ' has thus played a significant part in the development of veterinary science.

One further service of the ' Royal ' must be shortly recorded here. The farmers of other countries have long been accustomed to come to Britain for high-class live stock, and the export trade in pedigree animals has long been important to British Agriculture. In spite of our comparative freedom from epidemic disease, this trade has been liable to serious interruption through sporadic outbreaks of foot-and-mouth disease, most importing countries having taken the view that imports could be allowed only when the exporting country had a perfectly clean bill of health. The only solution of the difficulty, as it seemed to the Council of the Society, was to establish a Quarantine Station. After a good deal of discussion and negotiation with the Ministry, it appeared that the most desirable plan would be that the Society should itself undertake to establish the Station and to run it for a trial period. This responsibility was accepted, the Empire Marketing Board providing the capital required, while the Ministry undertook to arrange for the daily inspection of the quarantined stock and to issue the necessary health certificates.

The Station was constructed at London Docks, and was opened in 1928. It operated with perfect success, though at some expense to the Empire Marketing Board, until 1933, when the Board's activities ceased. The quarantine fees were then raised to such a level as could be expected to make the Station self-supporting, and in March 1934 the control was handed over to the Ministry of Agriculture. During the six years of the Society's responsibility a total of 2,696 animals passed through the quarantine and, apart from one minor case of mange, there was no outbreak of parasitic or contagious disease.

CHAPTER IX

AGRICULTURAL SCIENCE

THE Society, at various periods in its history, has used a variety of means for the promotion of agricultural research and for the dissemination of scientific knowledge among farmers. As is shown in a later chapter, the *Journal* has always played its part in making known research work that seemed likely to find an application in practice ; indeed, before the days of publications devoted to this particular purpose, the *Journal* was the medium chiefly used ; the bulk of the early Rothamsted work, for instance, is reported in the volumes of the 'fifties to the 'nineties. Another means, largely used in the Society's early years, was the arrangement of lectures by leading scientific men, either to the monthly meetings or to the general meetings of members.

One of the earliest actions of the Council was to appoint a consulting chemist and, in course of time, four scientific officers, besides the consulting engineer, came to be employed—a chemist, a veterinarian, a botanist and a zoologist. These officers had two functions. The one was to provide a scientific advisory service for individual members, whereby they could get analyses of soils, fertilisers, feeding stuffs, water and agricultural products, could have seeds examined for purity and germination, could have weeds and grasses identified and could obtain advice about the control of plant and animal diseases and pests. Up till the time of the appointment of County Agricultural Organisers and County Analysts, and the institution of State-aided Advisory Centres, the Society's scientific advisory service was almost the only one that was available. Incidentally, the scientific officers, and the committees under whom they worked, accumulated a unique knowledge of the problems with which they were concerned, and this knowledge was of great value when the time came for such legislation as the Fertilisers and Feeding Stuffs Act, the Seeds Act and the Agricultural Holdings Acts. The second function of the scientific officers has been to carry out research.

Again, the Society has given a substantial amount of financial help to various research institutions at times when these were in need of funds for expansion. Rothamsted, which almost from the time of its foundation had the closest relationship with the ' Royal,' has more than once been generously supported ; Cambridge University, and the National Dairy Research Institute at Reading, are among other bodies that have benefited.

The Society's largest venture into the field of research was to run, over a period of forty-four years, the Experimental Station at Woburn ; the results achieved are described below. The connexion with Woburn was severed in 1921, but the Station is still carried on under the ægis of Rothamsted.

When Woburn was given up, the Council decided to establish a Research Committee and to place at its disposal funds which, in most years, have amounted to about £2,000. These funds have been used, as to the lesser part, for the production of an annual review of agricultural research (*The Farmer's Guide*) ; and, as to the greater, to provide grants for such investigations as the Committee has selected.

This chapter traces, more or less in chronological order, the development of these various activities.

In 1847 Dr. Lyon Playfair felt obliged, owing to the increasing volume of the work, to give up the post of Consulting Chemist, and J. T. Way was appointed in his stead. Way was the first professor of Chemistry at the Royal Agricultural College, Cirencester. For some little time he carried on both appointments, but in 1849 he came to London and set up his laboratory in Holles Street, Cavendish Square. In the same year the Chemical Committee drew up a new and detailed scale of fees for chemical analyses and reports made for members of the Society. Way was paid a retainer of £200 a year and received the fees that members paid for his services. The Committee had a further annual grant of £300 for chemical enquiries, and some part of this generally went to meet the cost of special investigations which Way himself undertook ; one such task that he carried out was a detailed analysis of the ash constituents of all the more important crop plants, the results of which threw a good deal of light on their varied requirements in the way of mineral nutrients.

Way was a man of very high scientific attainments and he carried out, during his time with the Society, a large amount of original research. His special interest was in the fundamental properties of the soil—a subject to which, rather curiously, Lawes was never attracted.

In 1849 Way's attention was directed to one of the most interesting properties of soil by some experiments carried out by H. S. Thompson (Sir Harry Meysey Thompson). Papers both by Thompson and by Way were printed in the *Journal* for 1850. Thompson had been doing some chemical research in collaboration with Joseph Spence, a chemist and druggist in York, and one of their experiments was to fill a long glass tube with soil, to pour solutions in at the top and to analyse the drainage water that passed out at the bottom. The curious fact was discovered that, when a solution of ammonia was poured in, the soil somehow " absorbed and assimilated " the ammonia, and retained this even when the contents of the tube were leached with pure water. Way was quick to see the interest of this discovery, and he actively exploited the method of experiment that Thompson and Spence had used. He found that not only ammonia, but also potash and phosphates, when applied to the soil in soluble forms, were precipitated, and could afterwards be extracted only by the use of strong chemicals. This explained a good deal—in particular the failure of Liebig's patent manure ; the latter was composed of potash and phosphate, but these were fused with lime in the process of manufacture in order, as Liebig thought, to ensure against the risk of their being washed out by rain ; the fact was that the fusion process was not only unnecessary, but rendered the manures so insoluble that they were not available to the plant.

Two years later, still following the same track, Way got very near to the modern conception of the constitution of soil clay, upon which the absorptive powers of the soil depend. Again, in 1856, he discovered one of the most important facts about the nitrogenous constituents of the soil—namely, that nitrates are formed, in the soil, from other nitrogen compounds. Unfortunately he failed to realise the importance of this discovery and continued to believe that it was ammonia, rather than nitrate, that the plant absorbed. All this work was of prime importance, so that Way is now recognised as one of the founders of soil science.

In 1857 Way resigned to take up a public appointment and was succeeded by Augustus Voelcker, who had already followed him as Professor of Chemistry at Cirencester. Voelcker was a German who had begun his career, at the age of 16, as assistant to a pharmaceutical chemist in his native town of Frankfort-on-Main. Six years later he went to Göttingen and studied under Wöhler, then probably the greatest teacher of chemistry in any country ; he also went to Giessen, on purpose to hear Liebig's lectures, and here he met Henry Gilbert, who was later to be Lawes' colleague at Rothamsted. The two formed a close friendship which lasted till Voelcker's death.

In 1847 Professor Johnston, who was consulting chemist to the Highland Society, discovered Voelcker working as a research assistant at Utrecht and persuaded him to move to Edinburgh as his assistant. Voelcker's knowledge of farm practice, which was remarkably sound, was first gained through his contacts with Scottish farmers. He went to Cirencester in 1849 and there, besides his teaching, he got through an immense amount of experimental and analytical work. He published a great deal on Agricultural Chemistry and showed, for a foreigner, a remarkable command of English.

Under Voelcker the volume of the Society's chemical work steadily increased, and in 1863 he decided to give up his chair at Cirencester, to set up his laboratory in London and to devote himself to the work of the Society and to his growing practice as a consulting chemist. In 1865 more than 300 members' samples passed through his laboratory and by 1884, the year of his death, the number had risen to more than 1600.

The analytical work of his laboratory reached so high a standard of accuracy that he became the recognised court of appeal on agricultural analysis. Moreover, Voelcker was absolutely fearless in his dealings with manufacturers and merchants, and was no respecter of persons ; he exposed adulteration and every kind of fraud wherever he found it, and thus did a great deal to raise the standard of integrity in the fertilisers and feeding-stuffs trades. The reputable merchants (who were not a few) welcomed his activities wholeheartedly.

In 1869 the Council gave consideration to the unsatisfactory state of affairs in these trades and decided, for the future, to give the widest publicity to Voelcker's reports and to publish the names of fraudulent persons in the public press. It was not to be expected that even Voelcker would be infallible, and it was realised that the Society might probably become involved in lawsuits ; but it was felt that this risk was one that should not be shirked.

The first lawsuit came in 1871. Among many other cases of fraudulent adulteration, Voelcker reported on a material which had been sold as bone dust but which was mixed with a large proportion of rubbish. Unfortunately the person named in the report was the manufacturer, who was able to show that the material, when it left his works, was genuine and that the blame belonged to an intermediary merchant. The Society acknowledged its error and got off with nominal damages.

The second action, in the following year, was a very involved one. The trial lasted three days, and the report of the proceedings filled two hundred pages of the *Journal*. A merchant had sold, as linseed cake, a material containing so much in the way of other ingredients that, in Voelcker's opinion, the description constituted a fraud. The trial turned on a question of definition ; it was held that the description of a cake as

' linseed ' could not, in law, be taken to imply a guarantee of any particular degree of purity ; but the Judge made some very frank comment on the subject of commercial honesty, and the jury awarded only ten guineas by way of damages. The case had the very desirable effect of causing a change in the practice of the trade ; the Hull seed-crushers and merchants decided that, for the future, " No other cakes than pure linseed cakes shall be sold as linseed cakes " and " that all mixed or compound cakes shall be described as such." The moral victory thus lay with the Society, and Voelcker's position, as a scourge for the ungodly, remained unimpaired.

Apart from his routine analytical work Voelcker carried out a long series of chemical investigations, and hardly a number of the *Journal* appeared without some useful article from his pen. He covered the whole range of agricultural chemistry, continuing Way's work on soils, collaborating with Rothamsted on the investigation of drainage waters, and investigating a wide range of agricultural products, from fertilisers to milk and cheese. Not the least of his services was the training up in his laboratory of a large number of young men who carried his knowledge and his standards of accuracy to all parts of the country. The value of his work was recognised by the scientific world in 1870, when he was elected a Fellow of the Royal Society.

In 1876, in collaboration with Lawes, he planned the original Woburn experiments. By 1879 his private laboratory had become inadequate to the amount of work that was coming in, and the Society provided another for his use, in Hanover Square. Voelcker died in 1884 and was succeeded by his son, J. A. Voelcker, who had already served a long apprenticeship in the family practice, and who inherited both the ability and the courage of his father.

In 1876, when the decision was taken to start the field experiments at Woburn, Rothamsted was still the only agricultural experimental station in England. The classical experiments on the continuous growing of wheat and barley were exciting a great deal of interest. It appeared that, in fact, crops could be grown on, and removed from, a given piece of land for an indefinite period, and that a very fair level of production could be maintained by the use of chemical fertilisers alone. It seemed very desirable to repeat the experiments on a different type of soil, in order to find out whether the chemical theory of plant nutrition was of general application.

This was one, though not the chief object of the new station ; the more immediate object was to test, by means of field experiments, the actual manurial values of feeding stuffs consumed on the farm.

SIR JOHN BENNET LAWES, 1814–1900
From a painting by Hubert Herkomer, R.A

SIR HENRY GILBERT, 1817–1901

Augustus Voelcker, F.R.S.
CONSULTING CHEMIST, 1857–84

Dr. J. A. Voelcker
CONSULTING CHEMIST, 1884–1937

It was already an old custom, in many parts of the country, to award some compensation to an outgoing tenant for the presumed value, as manures, of feeding stuffs bought, and consumed by live stock, during the last years of the tenancy. This might, in the case of those consumed in the final year, be calculated at half or one-third of the price paid. Lawes and Gilbert had given some attention to the basis upon which residual values should be calculated, and had made the point that the price of a feeding stuff might bear little relation to the value of its residues. In 1870 Lawes read a paper on the problem to the Farmers' Club and, five years later, published in the Society's *Journal* a long article on the subject. The latter included a table of provisional residual values, based upon the composition of the individual feeding stuffs. The general opinion among valuers was that Lawes' values were too high, and subsequent work was to show that the critics were right. The question, however, was not one that could be settled by argument ; field trials were clearly required and with the passing, in 1875, of the first Agricultural Holdings Act the need for a basis of actual fact became more urgent.

In the Autumn of 1875 Mr. C. Randall proposed in Council that the Chemical Committee be asked to consider whether it was desirable and possible to carry out a series of experiments on the subject ; he thought that these might be arranged on selected farms, in different parts of the country, under the supervision of Dr. Voelcker and the Chemical Committee. The Committee considered the proposal, agreed that the work was one which the Society might very usefully undertake, but thought that, if accurate results were to be obtained, the Committee itself would require to have complete control over the feeding of the stock, the disposal of the manure and the cropping of the land to which the dung was applied.

At this stage the Duke of Bedford very generously offered to provide a properly equipped farm and to bear the whole cost of the experiments, excepting only that of the scientific work which would be involved. The ninety-acre farm at Woburn which he first suggested was found to provide too small an area of sufficiently uniform soil, whereupon he asked the Committee to choose an additional area in the vicinity ; this was found in the now famous Stackyard Field. The Duke's offer was gratefully accepted, the special buildings were erected and the experiments were begun. Lawes did not long remain as Joint Director, but he and his successors at Rothamsted continued in consultation with the Voelckers during most of the time that Woburn remained under the Society's control.

Stackyard Field was well chosen to provide markedly different conditions from those at Rothamsted, for the light 'hungry' and naturally lime-deficient soil and the sandy subsoil of the one were in marked contrast to the chalky clay-loam of the other. In this trial two series of plots were laid out, and each received the same fertiliser treatment and carried the same crop (wheat or barley) every year. Some plots had single fertilisers, some combinations, one had dung and still another rape-dust as a second type of organic manure. Finally there were two unmanured plots as controls.

The original experiment to test the residual values of feeding stuffs was laid out with sixteen plots, in four sets of four ; each set constituted a miniature farm, run on the traditional four-course rotation of wheat, roots, barley and clover. For one of these farms the dung was produced by cattle receiving a nitrogen-rich feeding stuff, decorticated cotton cake. The second had dung produced with maize meal, a material whose analysis indicated a very low fertilising value. The remaining two were manured with dung produced from roots and straw only, but both had applications of artificial fertilisers ; these latter were calculated to supply, in the one case, the same amounts of plant nutrients as the cake and, in the other, the same amounts as the maize. The yields of roots, barley and wheat were carefully recorded. After an eight years' run the plan was changed in so far that the root crop was folded with sheep, the sheep getting cake, or maize, or no concentrate, and the lack of the concentrate being made up, as before, by the use of fertilisers.

In 1899 a second attack was made on the problem by investigating the losses that occurred in the making and storage of farmyard manure. In other words, these experiments were planned to determine, under conditions of good farming practice, how much of the fertilising constituents of a feeding stuff actually reached the soil.

As time went on the experimental programme was extended. In 1897 a series of green-manuring trials was laid down, the scheme being to grow a cereal crop (generally wheat) on the land in alternate years, and to maintain the humus content of the soil by growing and ploughing in green crops. The discovery of the power of leguminous plants to use the nitrogen of the air had recently been made, and it was one of the objects of the trial to compare the value of vetches, as green manure, with that of non-leguminous species (mustard and rape). The plots had dressings of phosphate, potash and lime, but no nitrogen fertilisers.

In 1897 a bequest of £10,000, under the will of Mr. E. H. Hills, enabled the Society to equip a field laboratory and to appoint a resident chemist. His chief duty was to conduct the pot experiments required to test the fertilising value of what the testator called " the rarer forms of plant ash " or what we should now call " trace elements " ; those particularly mentioned in the will were fluorine, manganese, iodine, bromine, titanium and lithium.

Among the more important of the remaining crop experiments were a series on the palatability, nutritive value and powers of persistence of pasture plants ; on seeds mixtures for pasture ; on the manuring of pasture ; on the comparative efficiency of the processes of ensilage and hay-making ; and on the control of plant diseases, especially potato Blight and finger-and-toe in swedes. Again, there was a long series of cattle-feeding and sheep-feeding trials which produced much valuable information. Perhaps the most useful of all were those on calf-rearing, which demonstrated the possibility of rearing calves on separated milk with a cheap butter-fat substitute in the form of oats.

It is impossible here to discuss, in detail, the outcome of all this work, a full account of which, by the late Dr. J. A. Voelcker and Sir E. John Russell, was published in 1936.[1] Some of the conclusions must, however, be shortly noted.

Broadly speaking, the continuous wheat and continuous barley experiments confirmed the Rothamsted results ; they showed that, in either case, yields could be well maintained over a long period of years by the use of " complete " fertiliser dressings, i.e. of adequate amounts of phosphates, potash, nitrogen and lime. The striking difference arose from the difference in the lime reserves of the Rothamsted and Woburn soils. At Woburn the continued use of sulphate of ammonia caused a serious lime deficiency and, in the end, brought about such a high degree of acidity that the plots in question grew almost nothing except spurrey. The condition, of course, could be remedied by applying lime. The trials provided very valuable information on the rate of loss of lime from light soils, and on the influence of different fertilisers upon the amount of this loss.

The first eight years' work on the residual values of feeding stuffs fed to cattle produced very surprising and disconcerting results. The yields were as shown in table on page 126, the manures for the whole rotation being applied, in each case, to the root crop.

The difference between cake-fed and maize-fed dung was almost negligible, the former giving a slightly heavier yield in the year of applicacation but the latter showing, perhaps, a rather longer-lasting effect. Poor-quality dung, supplemented with artificial fertilisers, had a substantially better effect than cake-fed dung. The experiment may be said

[1] *Fifty Years of Field Experiments at the Woburn Experimental Farm* (Longmans Green).

Kind of dung applied . .	Cake-fed	Maize-fed	No Concentrates Fed	
Artificial fertilisers . . .	Nil	Nil	Equivalent to Cake	Equivalent to Maize
Yields of crops :				
Roots (tons per acre) .	14·39	13·75	17·78	14·94
Barley grain (cwt. per acre)	23·0	22·6	24·2	21·6
Wheat grain (cwt. per acre)	22·6	23·6	22·8	23·0

to have failed to show any appreciable result, as regards the fertility of the land, from the feeding of a cake which had, in theory, a high residual manurial value.

In the second series of trials the cake and maize were fed to sheep folded on the root crops, and the artificials were applied to the succeeding barley ; but here again the effect of the cake was slight and short-lived, and was substantially less than that produced by an " equivalent " dressing of artificial fertiliser.

It must be said that dissimilar results have been obtained from similar experiments at other centres ; at Rothamsted, for instance, the residues from cake feeding have produced substantial crop responses, though here, again, the effects have been very short-lived. The failure of the Woburn soil to respond to cake residues is at least partly to be explained by its nature. In well-aerated sandy soil the nitrogen compounds in these residues are very quickly converted into nitrate which, in turn, is very readily washed down into the drains ; probably, under the conditions at Woburn, a large proportion was lost before the plant roots were ready to absorb it.

It was a disappointment to all concerned that this problem, the one with which the station was mainly concerned, was left without any complete answer ; the only conclusion was that, under certain conditions, residual values of feeding stuffs could be much less than had been generally supposed.

Turning to the investigation of losses in the dung itself, the results may be put very shortly. Cattle were fed indoors on normal rations and the foods were weighed and analysed for nitrogen ; the animals were weighed at the beginning and end of the feeding period and the total nitrogen in their live-weight increase was estimated. The difference was thus, presumably, the amount of nitrogen voided in the manure. The dung produced (in concrete-floored boxes) was weighed and analysed at a time when it was carted out in early Winter ; at this stage the loss was found to have reached about 15 per cent. The manure was then drawn out to the field, built into a compact heap and covered with earth ; it was again weighed and analysed in the Spring, when it was due to be applied to the land. At this stage the loss had reached 33 per cent. Thus, under the best conditions likely to prevail in practice, only two-thirds of the nitrogen voided by the animal actually reached the soil. These results were used by Hall and Voelcker, in 1901, in preparing a revised table of residual manurial values.

The green-manuring experiments provided another surprise. In theory, vetches should be a better green manure than mustard or rape, for by their growth they enrich the soil in nitrogen. The estimated average addition from the crops grown at Woburn was about 40 lb. of nitrogen (equivalent to fully 2½ cwt. of nitrate of soda) per acre. A Summer crop of rape or mustard will, of course, conserve nitrogen that would otherwise be lost, but it adds no nitrogen to the soil. It was thus to be expected that the vetches would prove the better preparation for corn, and, year by year, when the plots were inspected in Spring, this expectation seemed to be borne out ; the corn after the vetches looked the better ; but year by year, with few exceptions, the position was reversed before harvest, the crop following the vetches failing to fulfil its early promise. The average yields of the fourteen corn crops, grown between 1893 and 1917, were as follows :

	After Vetches	After Mustard
Grain (cwt. per acre)	10·5	14·2
Straw (cwt. per acre)	14·9	19·8

These trials were analysed by E. M. Crowther and H. H. Mann in the Society's *Journal* for 1933, when the results, which had puzzled a generation of farmers and scientists, were at last explained. The fact is that the material in a vetch crop, because of its richness in nitrogen, behaves, in a light soil, as a very quick-acting manure, being largely broken down and nitrified within a few months of its being ploughed in. Hence the two successive vetch crops ploughed under (as was the normal practice) in June and September, would yield up nitrate chiefly in the Autumn and Winter, and there would be very little left for the corn plant to feed on during its main feeding period. Mustard, on the other hand, because it is poorer in nitrogen, breaks down much more slowly, and yields up nitrates chiefly at the time when these are of use to the succeeding crop.

This experiment, taken in conjunction with that on the residual value of cake, provides an important lesson in connexion with the management of light land—that a "hungry" soil must be fed, but that the great speed with which it digests its food necessitates a careful timing of the applications.

The story of the pot experiments carried out under the Hills Bequest is the rather tragic one of a sound idea that was produced before its time. We now know that many "trace elements" are essential for the nutrition of the plant ; we know that boron deficiency causes disease in sugar beet and swedes ; that a shortage of available manganese in the soil leads to failures of oats ; and that zinc, copper, and many other elements are required for normal plant growth. But the amounts that plants require are extremely small ; 10 lb. of borax, applied to 1,500 tons of soil, will provide all that a crop of swedes or sugar beet requires. Hence experiments on the effects of these substances upon the plant require the most elaborate precautions if their complete absence, in the case of the ' no manure ' controls, is to be assured. At the time of the Woburn experiments chemical technique was not capable of ensuring this, and no doubt traces of the various elements were present in cases where their absence was assumed. It is not surprising that the results were largely negative. Nevertheless, the experience gained was to prove of value to later workers.

The successive Dukes of Bedford continued to bear the maintenance costs of Woburn until 1913, and from this date until 1921 the Station remained the responsibility of the Society, with the help of a grant of £500 from the Board of Agriculture.

We must now turn from chemical to biological science. The appointment of a consulting biologist was a comparatively late development, but the Society, from time to time, took other steps for the encouragement of biological research. Their efforts in connexion with potato disease may be mentioned as an example. It is well known that the outbreak of potato Blight in 1845, and the almost universal and disastrous epidemic of 1846, produced a crisis in the country. Where, as in Ireland and the West Highlands of Scotland, potatoes had become the main food of the people there was severe famine—the last that our country experienced. The population of Ireland was reduced, by death and emigration, by more than two million people. The Society, like all other bodies who were concerned with agriculture, did what it could by publishing every suggestion that seemed to hold out any hope of a mitigation of the losses. Some degree of improvement was secured by the production of new seedlings, notably Paterson's Victoria, bred about 1855 and put out for general cultivation some six years later. But this was only a partial solution of the problem.

1872 was again a year of widespread outbreaks and in 1873 Earl Cathcart, the Society's President, offered a hundred-pound prize for the best essay on the subject. Ninety-four essays were submitted, but the Committee which was appointed to adjudicate was not able to recommend an award. They suggested that a grant should be made to a competent mycologist to investigate the life history of the fungus and especially to find out how this survived over the Winter ; and, further, that the Council should offer prizes for disease-resistant sorts. Professor De Bary, of Strasburg, was appointed to carry out the scientific work and his paper on the fungus appeared in the *Journal* for 1875. This did not lead to the discovery of any infallible preventive, which indeed is still to seek, but it laid a valuable foundation of real knowledge about the Blight fungus.

In 1871 the Council decided to appoint a Consulting Botanist, whose duties would be to examine seeds and plants for members, and to write for the *Journal* papers on subjects of botanical interest. Their choice fell on William Carruthers, who had recently been appointed chief of the Botanical section of the British Museum and had just been elected a Fellow of the Royal Society in recognition of his research work.

At the time of Carruthers' appointment the trade in seeds seems to have been in an even worse state than that in fertilisers and feeding stuffs. Cereal and root seeds, in fact, were usually of satisfactory quality, but clovers and grasses were frequently bad—full of impurities, often deliberately adulterated and sometimes so treated, in order to improve their appearance, that the germinating capacity might be reduced to zero. By 1877 the Botanical Committee felt sufficiently sure of its ground to ask Council for the same kind of powers that had already been given to the Chemical Committee, i.e. the authority to publish the names of persons who had sold to members of the Society seeds which had been determined by the Botanist to have been killed, artificially coloured or adulterated. The reputable seed-merchants welcomed the step, and its effects were highly salutary.

In 1882, after ten years' experience, the Committee felt sufficiently sure of its ground to lay down standards of purity and germination for all the commoner farm seeds, and recommended that farmers should insist upon guarantees with all seeds that they bought. The result was that, in the following year, two of the largest seed firms gave guarantees, up to or above the Society's standards, with all the seeds that they sent out. In the immediately succeeding years many other seedsmen felt obliged to follow suit. In the end, a Seeds Act was necessary to curb the activities of a small minority of merchants, but it would be difficult to overestimate the benefits secured through the Botanical Committee's efforts.

The next subject to engage the Committee's attention was the choice of species for sowing down pastures, more particularly long-duration leys and permanent grass. In the 'eighties, the mixtures commonly used consisted very largely of perennial and Italian ryegrasses, with red, alsike and Dutch white clovers. The only ryegrass available was, of course, of the " commercial " or long-cultivated type, and the value of wild white clover was as yet not realised. This old type of mixture quickly produced a sward, but this, on all but the best land, very quickly deteriorated. Carruthers carried out a systematic series of botanical analyses of first-class pastures in many parts of England and thus got a much clearer idea of what should be the aim in sowing down. At Woburn, and also on the farm of Mr. Faunce de Laune, who co-operated in the work, numbers of plots were sown with pure species in order to obtain information about the yield, palatability and powers of persistence of each. This work suggested the elimination from mixtures of several species whose use had formerly been recommended. The general conclusion from these trials was that cocksfoot, timothy, meadow fescue and foxtail, with red and white clovers, should form the basis of mixtures for permanent grass. The omission of ryegrass is interesting, and there is no doubt that its complete condemnation was a mistake. We must remember, however, that the difference between ' commercial ' and indigenous strains was not appreciated at the time in question, and we can easily believe that the commercial strains, sown without wild white clover, would generally have a very short life.

Another very interesting piece of work was an experiment, laid down in 1895, on the much disputed question of the longevity of seeds ; some of the results are still worth quoting. Seeds of forty-three different species and varieties of crop plants were stored in paper bags, under ordinary indoor conditions, and samples were drawn from each, year by year, and germinated. The full report appears in the *Journal* for 1911. The successive percentage germinations of a few of the sorts, at intervals of three years, are shown in the table on opposite page.

The paper gives a great deal of interesting information about the vigour of germination, and was of value as showing how long the seeds of the various species could be stored without material loss of value.

Another important branch of the Botanist's work was the investigation of plant disease. Carruthers and his assistant Güssow succeeded in finding

	Barley	Red Wheat	Black Oats	Perennial Ryegrass	Red Clover	White Clover	Swedes
1896 . .	99	99	97	95	98	99	100
1899 . .	95	88	94	81	95	84	92
1902 . .	25	79	92	36	7	48	84
1905 . .	0	0	88	10	2	7	14
1908 . .	—	—	34	0	0	0	0
1911 . .	—	—	0	—	—	—	—

the cause of the common and most important form of " clover sickness." Again, as has been mentioned above, the Society offered prizes for Blight-resistant varieties of potatoes, and although no variety stood up to the test which was applied, Carruthers' reports on the twenty trials gave the farmer detailed information about the relative disease-resisting powers of the varieties entered. The experiments on Bordeaux Mixture, at Woburn, showed that this material was a valuable preventive of Blight.

Carruthers retired in 1908 and was succeeded by Professor R. H. Biffen (Sir Rowland Biffen), whose services to British Agriculture are well known. What is perhaps less generally realised is that the methods of wheat breeding which he has developed have been of immense value to the world at large ; they have been most successfully applied to the production of drought-resistant forms for arid regions, of early-ripening types for areas with short Summers, of types of high baking quality and of others with remarkable disease-resisting capacity.

The staff of scientific consultants was completed, in 1882, with the appointment of Miss Eleanor Ormerod as Honorary Consulting Entomologist. Miss Ormerod had taken up the study of insects as a hobby, and gradually became interested in these from the point of view of their importance as farm, garden and woodland pests. In 1877, in conjunction with her sister (who was a very competent artist), she began the issue of an 'Annual Review of Observations on Injurious Insects,' which she published, at her own expense, throughout the remaining twenty-four years of her life. On the day of her appointment by the ' Royal ' she met with an accident which left her permanently lame, but this disability was not allowed to interfere with her work, and she proved a most competent adviser to the Society's members. Towards the end of her life she was given an Honorary Doctorate of Edinburgh University—the first woman to be so honoured. Miss Ormerod continued as Consulting Entomologist until 1892, and was succeeded by Mr. Cecil Warburton of

Cambridge University, who has since maintained and extended the service which she built up.

A new if minor step towards the encouragement of research was taken in 1911, when the Council decided to offer annually a gold medal for research in Agriculture, the award carrying with it the Life Membership of the Society. The age limit of competition was fixed at twenty-seven, and special judges were appointed, each year, according to the subjects of the papers that were submitted. The paper of the successful candidate was to be published in the *Journal* if it was considered suitable for that purpose.

The scheme, over the following period of twenty years, produced some useful contributions to knowledge, but, upon the whole, it was not as successful as had been hoped. In 1933 the Council decided to discontinue the offer, and substituted a new medal to be awarded for distinguished services to Agriculture (see Chapter XIII).

The last important development in the field of research was the setting up, in 1921, of the Research Committee. This step was related to the giving up of the Woburn Station. The Dukes of Bedford had borne the cost of the Woburn work until 1910 but, with the establishment of the Development Commission and the provision of State funds for agricultural research, it seemed reasonable that the cost should no longer fall upon a private individual. The Council applied to the Commission for a grant-in-aid, and in 1913, after long negotiations, a sum of £500 a year was obtained. A great deal of work was going on at Woburn at the beginning of the War, but the depletion of the staff, and the necessity to produce the maximum of food from the land, led to the rapid curtailment of the experimental programme, so that, by 1918, little was being done except to maintain the long-period experiments.

In 1920 there was a minor crisis in the financial affairs of the Society. Darlington Show had produced a heavy loss, while a high general price level was causing an alarming increase in general expenditure, and there was no compensating rise in the ordinary revenue. A special (economy) Committee was appointed, and one of its recommendations was that the Woburn experiments be discontinued. There was a party opposed to the step, but the recommendation was carried by a large majority of the Council ; the work under the Hills Bequest was transferred to the School of Agriculture of Cambridge University ; the field experiments at Woburn were carried on by Dr. Voelcker, personally, for the following five years, after which time the Station was taken under the wing of Rothamsted.

The Society, however, was still anxious to do its part for the promotion of Research, and an unexpectedly large improvement in its finances, in 1922, made possible a consideration of the various possibilities. The Finance Committee thought that £2,000 could safely be allocated to this object, and that annual grants of about this amount might be possible for the future. A special Committee was appointed to consider how the scientific side of the Society could be developed. They reported that the results of past experimental work should be collated, abstracted and published ; that members of the Society be asked to make suggestions about practical problems that seemed to require investigation ; that research institutions be asked to undertake investigations which the Society would finance ; and that the administration of the fund, and the direction of the experiments, be placed in the hands of a permanent Research Committee.

The work carried out under the Committee has been reported annually in the *Journals* from 1924 onwards ; it is here possible to mention only a few of the more interesting results.

One item in the first programme was a series of investigations on the utilisation of whey. Work on condensation, already being done by the Ministry of Agriculture, was supplemented by experiments carried out under the Society at Reading, and feeding trials were arranged in various parts of the country. The latter showed that whey, used as a relatively small part of the rations of pigs, had a very favourable effect upon their growth and general thriftiness. Also a new method of extracting lactose (milk sugar) from whey was worked out by Dr. Leonard Harding.

Another piece of research which has been maintained almost continuously throughout the Committee's existence is that on the inoculation of leguminous plants with strains of nodule-forming bacteria. This work, carried out in the Bacteriological Department of Rothamsted, soon led to the isolation of a really effective strain of the lucerne organism, and the use of this, for the inoculation of lucerne seed, has now become standard practice. The investigations have more recently turned to clover, and the isolation of the best strains of bacteria for white clover promises to produce valuable results, at least in certain localities where the prevailing strains in the soil are ineffective.

Grants to the Norfolk Agricultural Station enabled a valuable series of bullock-feeding experiments to be carried out. The first important point to be established was that satisfactory fattening could be secured with much lower allowances of protein, and hence of expensive cakes, than farmers had been accustomed to use. The net saving achieved by the use of the new type of ration (which had originally been suggested by the late

Professor T. B. Wood) amounted to some 30s. per head during a feeding period of ordinary duration.

Other very useful information was derived from the Norfolk Station's work on the utilisation of sugar-beet by-products (tops and wet and dried pulp), a problem that arose with the extended cultivation of the beet crop under the Sugar Subsidy. Here it was shown that beet tops, used with certain precautions, could replace swedes on a weight-for-weight basis : that one pound of dried beet pulp could fully replace seven pounds of swedes ; and that wet beet pulp, either fresh or ensiled, could be largely used in Winter cattle feeding.

Early in the Committee's life a fresh attack was made on the old problem of residual values by laying out a trial on the value of cake feeding as a means of grassland improvement, but here again the results of the trial were almost completely negative. A long-term trial, laid down at Rothamsted, has not yet reached the stage when results can be expected.

A series of experiments on the cultivation of malting barley, again carried out at the Norfolk Station, produced several valuable items of information for growers. Of these perhaps the most striking was that the new hybrid barleys (Plumage-Archer and Spratt-Archer) would bear and amply repay much more generous manuring than the sorts which had been widely cultivated in previous times.

The Research Committee also financed a long investigation on the control of mastitis in dairy cattle, the work being done by the Animal Pathology Institute of the Royal Veterinary College. The work showed that the eradication of mastitis, by methods of segregation, was not, as had been rather generally supposed, a hopeless undertaking, and steady progress is now being made towards a solution.

The Committee made provision, according to its original terms of reference, for the collation and statistical examination of the records of the Woburn experiments, and these were embodied in the book which has been referred to in an earlier part of this chapter.

Among the long-term experiments now in progress, apart from the Rothamsted experiments on the feeding of cake on pasture, is one on the value of various alternative methods of maintaining the fertility of light arable land. A further undertaking has been to arrange, in various counties throughout England and under a wide variety of climatic and soil conditions, trials of the pedigree indigenous strains of grasses which have been selected by the Welsh Plant Breeding Station under Sir George Stapledon.

This list is by no means exhaustive, yet it would seem to show that the

Society, with the rather limited means at its disposal, has enabled a considerable amount of very useful work to be done.

Another of the Research Committee's undertakings dates from 1925. Since that time it has arranged for the annual publication of a review of agricultural research, covering the subjects of Soils and Manures ; Crops and Plant Breeding ; the Feeding of Livestock ; Implements and Machinery ; Dairying ; Animal Diseases ; and Farm Economics ; with articles at longer intervals on insect and other pests. This publication has been welcomed by farmers, Agricultural Organisers and all those others who must try to keep themselves informed about scientific progress in all its bearings upon the industry but who cannot hope, themselves, to work through the vast and growing volume of scientific publications.

CHAPTER X

AGRICULTURAL EDUCATION

IT will be remembered that one of the original objects of the Society, as drawn up by the Provisional Committee, was "To take measures to improve the education of those who depend upon the soil for their support." In 1838 provision for agricultural education was of the most meagre kind. In 1790 a Chair of Agriculture and Rural Economy had been established at Edinburgh, and in 1796 an endowment had been provided for a part-time professorship at Oxford. The occupants of these posts at the time of the Society's foundation—Professor David Low and Professor Charles Daubeny—were both men of distinction and were recognised as two of the leading agricultural writers of the time ; but neither had more than a handful of students. Daubeny took an active interest in the young Society, was a frequent lecturer at its early meetings and was one of the first to receive its Honorary Membership.

The first important development, after 1838, was the foundation of the Royal Agricultural College. The movement to open a college was started by R. J. Brown in 1842 and, after two years of strenuous propaganda on his part, a meeting, which was held in Southampton at the time of the Society's Show, decided to proceed with the enterprise. The Society, as such, did nothing to forward the proposal, but Pusey took the chair at the Southampton meeting and was supported by Earl Spencer, the Duke of Richmond, Daubeny, Dr. Lyon Playfair and many other prominent members. Lord Bathurst offered a site for a College, as well as a long lease of a large farm, on his estate near Cirencester, a sum of £12,000 was raised by subscription, and the College obtained a Royal Charter in 1845.

Brown seems to have been disappointed that the Society should have refused its support to the College—especially when, in its early years, it got into rather desperate financial straits and indeed came near to closing its doors. But the students of the College were mostly the sons of landowners and other well-to-do people, and the Council evidently took the view that there were other causes more deserving of the Society's support. Nearly twenty more years passed before the Council seriously considered the problem of education. The matter was raised at the General Meeting of December 1863 by Chalmers Morton, the eminent editor of the *Agricultural Gazette*, who had, at one time and another, done a good deal of work for the Society. Morton said that, while the Society was doubtless doing a great deal for education in the general sense, it had as yet done nothing to discharge the particular duty which its founders evidently had in mind—*i.e.* to promote the systematic education in agriculture of the young generation. In the following April the subject was pursued by Edward Holland, M.P., a member of the Council, who moved for a committee to examine the whole question. A committee was duly appointed with Holland himself as chairman and with Wren Hoskyns, Colonel Kingscote, H. S. Thompson and others as members. The Council also put agricultural education on its list of subjects for Prize Essays, and all three of the essays which were submitted appeared in the *Journal* for 1866. Chalmers Morton and Holland, also, read papers on Education at Council meetings.

There seems to have been an almost unanimous opinion among members of the Council that something ought to be done ; the difficulty was to find something that the Society could usefully do with the sum that it could afford to spend. The Committee duly reported, its views were fully debated by the Council and the report was sent back for further information and reconsideration. The second report was not unanimous, but it was decided to give a trial to the scheme which was favoured by the majority. The object of this scheme was to encourage the study of science, in its bearings on agriculture, in higher schools, and the method proposed was to provide an annual sum of £200 to be awarded in prizes for proficiency in science subjects. The work of examining was delegated to the Oxford and Cambridge Local Examination Boards.

It is unnecessary to discuss this scheme in detail, because it proved an almost complete failure. In 1866 only seven candidates appeared, and in the following year the examination was cancelled on account of the meagreness of the entry. A new scheme was drafted on the general lines that had been suggested by the minority of the Committee. By this the Society set up its own examination, intended for students of more mature years, and appointed its own examiners.

Candidates were required to offer a minimum of four subjects— Practical Agriculture, Chemistry, Book-keeping and either Land Surveying or Agricultural Mechanics. They might offer also the second of these two optional subjects, with one or more from another list which at first included Botany, Geology and Anatomy and, later on, Agricultural Entomology and Veterinary Science. The examiners were instructed to award first-class and second-class certificates, and the first class was to carry with it the Life Membership of the Society. Further, the Society offered prizes of £25, £15, £10 and £5 to the candidates who should gain the first four places on the list, as determined by their total marks. Gold and silver medals were later substituted for these money prizes. The examination, as judged by the question papers published year by year in the *Journal*, was a searching one, extending over a period of five days.

For many years this enterprise achieved but small success ; up till 1882 only two institutions—the Royal Agricultural College and the University of Edinburgh—provided anything like an adequate course of instruction in the subjects demanded, and although, in that year, a lectureship in Agriculture was established in the Normal School of Science, South Kensington, the number of candidates remained small. During the first twenty years the number fluctuated between two and twenty-one, and showed no distinct upward trend. Moreover, the standard of attainment, upon the whole, seems to have been low, for the average number of certificates awarded was less than four per year. The students of that generation seem to have shared the farmer's proverbial distaste for accounts ; at any rate the examiners had often to remark on the high proportion of miserable failures in book-keeping.

During all this time many members of Council doubted whether the results of the scheme justified the expenditure of the two or three hundred pounds that were in fact spent on it each year. But nobody pressed the point and the examination continued to be held. After 1890, as is shown below, conditions changed, and the examination began to serve a real purpose.

A second examination was established in 1874. By this time a number of Middle Class schools were teaching agricultural science, and the headmasters of these wanted a standard form of examination, as well as some means to encourage their pupils to proceed to more advanced study. After a conference between the headmasters and the Society's education committee, it was agreed to set up a Junior Examination, the subjects being Land Surveying, Mechanics, Chemistry and the Elements of Agriculture. At the same time the Society undertook to award annually, on the results of the examination, a maximum of ten scholarships of £20 each, it being a condition that the scholar should spend a year at an agricultural college or in the agricultural department of a University (Cirencester, Edinburgh or Glasnevin in Ireland) or, alternatively, as a pupil under a practical farmer or land agent to be approved by the Committee. Twelve schools joined in the scheme and the examination was held annually until 1895. The number of candidates varied from twenty to forty, and numbers of scholarships were awarded in each year, though only occasionally did the full ten candidates reach scholarship standard. The scheme came to an end in 1895 because, by that time, the County Councils were in a position to grant scholarships in all cases of need.

In 1881 a Royal Commission was appointed to enquire into "The instruction of the Industrial Classes of certain Foreign Countries in technical and other subjects, for the purpose of comparison with the corresponding classes in this country, and into the influence of such instruction on manufacturing and other industries at home and abroad." The whole enquiry into Agricultural Education was entrusted to H. M. Jenkins, the then Secretary of the Society. He was appointed to be a sub-commissioner, with instructions to report on the teaching of agriculture in France, Germany, Denmark, Holland and the United Kingdom.

Jenkins possessed all the qualifications for such a task, for he had travelled widely on the Continent, had written a good deal, in the *Journal*, on foreign agriculture, and was an accomplished linguist. His report occupies the greater part of the second Volume of the Commission's Report, which was issued in 1884. He described very fully the educational facilities provided in foreign countries, pointed to the contrast between the very meagre amount of State assistance provided in Britain and the much more generous provision elsewhere, and made many recommendations for the development of agricultural education in Great Britain and Ireland. The Commission adopted most of these recommendations, emphasised the special urgency of the matter in reference to Ireland, and added the suggestion that the National Agricultural Societies might perhaps divert some of their funds from the improvement of cattle and machines to the improvement of the minds of farmers' sons.

In 1888 a Departmental Committee reported in favour of State aid in the establishment of centres of agricultural education, and in the following year small grants were made to a number of County Councils and to the University College of North Wales, Bangor. By a fortunate accident a much more substantial endowment became available in 1890. In that year the Government introduced a Bill which had for its ultimate object the suppression of a proportion of the licensed houses in the country. A fund was raised, by means of licence duties, for the compensation of those holders who would be deprived of their licences. The Bill was, in fact, abandoned, but the fund remained and it occurred to somebody that the money could very usefully be devoted to the endowment of technical education. The Government adopted this suggestion. The "Whisky Money" amounted, in the earlier years, to nearly a million pounds a year, and of this some £80,000 came to agriculture. A good many County Councils consulted the Society on the method of expending their grants, and the Education Committee issued two reports making a number of suggestions, many of which were adopted. Besides Bangor, the Universities and University Colleges of Cambridge, Newcastle, Leeds, Reading and Aberystwyth were enabled, by the support of the County Councils, to form Departments of Agriculture, and the South Eastern (Wye), the Midland, and the Harper Adams Colleges were established by the same means. The expansion in teaching was reflected in the growing number of candidates for the Society's examination, which exceeded forty in each of the latter years of the century.

The development of County Classes after 1890 created the need for an elementary text-book of agriculture, and in 1891 the Education Committee asked the Council for leave to produce such a book. The general scheme was drawn up by a sub-committee, and the preparation of the book was placed in the hands of Dr. William Fream, who, two years later, became Editor of the *Journal. Fream's Elements* has since passed through twelve editions and has sold ninety thousand copies.

In the early 'nineties there was rapid expansion in the provision of instruction in dairying as well as in general agriculture. From the outset the Board of Agriculture had wished that some national examining body might be constituted in order to set a standard throughout the various teaching institutions, but the setting up of such a body was outside its statutory powers. The Board of Agriculture therefore proposed the formation of a Conjoint Examining Board, under whose direction all dairy examinations should be held ; and in 1895, with this object in view, they summoned a conference of representatives of the leading agricultural societies and of the various teaching institutions. Unfortunately no agreement could be reached, and therefore the Council of the ' Royal,' with the approval of the Board of Agriculture, decided to set up its own examination, and to award a Diploma to candidates who should " prove themselves capable of imparting instruction on the science and practice of dairying." The details of the syllabus were quickly worked out by the Education Committee, and the first examination was held at Reading in 1896, twenty candidates presenting themselves. Meanwhile negotiations had been begun with the Highland and Agricultural Society for the formation of a Joint Board to conduct the examination, under the same conditions, in both England and Scotland. These negotiations were soon concluded, the new Board met in April 1897 and the first examination for " The National Diploma in the Science and Practice of Dairying " took place in the following autumn. This examination, under regulations that have been revised from time to time, has been held in each subsequent year. The number of candidates has risen from 28 in 1897 to 65 in 1910, 131 in 1924 and 149 in 1938.

For many years before 1897 the Highland and Agricultural Society had been conducting its own Fellowship examination in agriculture, and this was similar in character to the senior examination of the ' Royal.' After a brief and entirely satisfactory experience of the joint arrangement for the dairy examination, the two Societies decided, in 1899, to join forces for the holding of a single examination in agriculture also. Accordingly, a National Agricultural Examination Board was constituted, composed of six representatives from each society. Regulations and a syllabus for the National Diploma in Agriculture were finally agreed in October 1899, and the first examination was held at Leeds in the spring of 1900. The syllabus covered both general and applied science as well ?s practical agriculture, and the ten subjects were divided into two groups, to be taken by candidates in successive years. A gold medal was provided for the candidate who should obtain the highest marks in the whole examination, and provision was made for the award of Honours. Fifty-one candidates, in all, were examined in 1900.

It seems unnecessary to trace in detail the considerable changes and developments in the examination during subsequent years. Their general effect has been to give it an increasingly practical bias. The Board now accepts the certificate of the Teaching Institution with respect to candidates' proficiency in the pure sciences, and increased emphasis is placed, in its own examination, on the subject of Farm Management. The intention has been to give the N.D.A. a specific character, and so to avoid the mere duplication of examinations.

The Board's examinations, both in Agriculture and in Dairying, have had their critics. It has been suggested that teaching institutions are better qualified than *ad hoc* examining boards to judge the standard of attainment of students ; that a University Degree or a College Diploma in agriculture provides all that the agricultural student requires ; and that the necessity to prepare students for an outside examination sometimes prevents the proper planning of their studies. But, whatever force there may be in these criticisms, there is no question about the sustained demand for both of the Board's examinations ; the increase in the number of entries for the National Diploma in Agriculture has been almost as marked as that indicated above for the Dairy qualification. The figures for 1910, 1924 and 1938 were respectively 93, 155 and 221.

Changes in the examination have sometimes originated from the Colleges and sometimes from the Board itself. In either case, however, there has been the fullest consultation between the Board and the teachers, and every effort has been made to meet the majority view of the latter.

CHAPTER XI

FARM PRIZE COMPETITION

ONE of the most valuable of the Society's activities was to run, from 1870 till 1892 and again from 1907 till 1915, a series of annual competitions for the best-managed farms in the areas where the country Meetings happened to be held.

The scheme originated in an offer by a private individual, made in connexion with the second Oxford Show of 1870 ; Mr. James Mason of Eynsham Hall, who was a very progressive local landowner, placed a hundred-guinea cup at the Society's disposal, to be awarded to the tenant of the best-managed farm in a defined area centred about Oxford. The Council gladly accepted the offer and undertook, for its own part, to add a second prize of £50, to bear the expenses involved, to draw up rules and conditions, and to appoint judges. Twenty-one entries were received.

It was part of the plan that the judges should draw up, for publication in the *Journal*, a reasoned report on the competition, with descriptions of the more noteworthy entries. The points to which they were directed to give particular attention were :

(1) General management with a view to profit.
(2) Productivity of the crops.
(3) Quality and suitability of live stock.
(4) Management of grassland.
(5) The state of fences, gates and roads and the general standard of neatness.

At later dates two other criteria were added, namely, the adequacy of the system of accounts and the length of time that the farm had been under the management of the entrant.

The Oxford competition aroused a great deal of interest and the judges' report presents a clear picture of the best farming of the district at a time when the combination of good prices with relatively low wages still provided a strong urge towards high production. The winning farm was one of some nine-hundred acres, on the shallow, brashy limestone land near Bicester, with over 800 acres of arable land farmed on the strict Norfolk four-course. There was, of course, a large folded flock of sheep (consisting in this case of 400 Lincoln ewes) whose produce was sold, after fattening on roots, as heavy tegs. Some seventy head of store cattle were wintered on straw with a small allowance of roots and a liberal ration of linseed cake. These were sold to graziers in spring, and were not expected to leave any direct profit—indeed, " if they paid the cost of the cake it was all that was expected of them." The only artificial fertiliser used was a light dressing of superphosphate to the roots, the fertility of the land being otherwise maintained by liberal cake feeding to both sheep and cattle and by the consumption, on the farm, of everything that was grown except the wheat and part of the barley. The farm thus exemplified the famous system of arable stock-and-corn farming which, originating in Norfolk a century earlier, had spread as far as the Yorkshire wolds, Shropshire and Dorset. The system enabled the farmer to pay a good rent for light and relatively poor land, left him with a good profit, provided, at what was then considered a reasonable wage, a large amount of employment and produced a high output of food. The judges made it clear that they had no bias in favour of this or any other system, but they thought that the Norfolk scheme was admirably suited to the conditions of this particular farm, and was being run to perfection.

It would, however, seem that, even in those days, some farmers were beginning to grudge the high expenditure associated with the arable-land sheep and yarded bullocks, and were seeking some less expensive way of fertilizing the land for corn. At any rate there was another corn-brash farm where two-thirds of the arable area (instead of only half) was under cereals, and where the farmer was pinning his faith to nitrate of soda and superphosphate as substitutes for the dung-cart and the sheep-fold ; but the judges thought that the substitution had been carried too far—" I know," says the reporter, " that Mr. Lawes contends that corn may be grown year after year by the use of artificial manures. I doubt, however, whether upon light thin soils the alternation of green and white crops can be profitably departed from. The inspection of twenty-one farms has impressed me strongly with the opinion that it can not." The general level of merit was so high that the judges asked for, and obtained, permission to award a third prize, and there were several other entries which won commendations. In reviewing the recent progress of the local farming the most striking improvement was in the quality of the sheep-stocks, more especially those of the new Oxford Down breed.

The Council was more than satisfied with the outcome of Mr. Mason's offer, and decided upon a similar but larger competition in connexion with the Wolverhampton Meeting of the following year. The local landowners provided two prizes, each of £100, for the best arable and the best dairy farm respectively ; the Society provided a second prize of £50 in each class and bore the incidental expenses. Entries of arable farms numbered 21 and the standard of merit was so high that two additional prizes were awarded. The dairy farms were only 7, but there was no difficulty in finding worthy recipients for the two prizes.

The Judges' description of the winning arable farm, on the Duke of Cleveland's Shropshire estate near Wellington, gives us some notion of the vast amount of capital that had been sunk in English soil during the prosperous days of the mid-nineteenth century. In this case, on a four-hundred-acre farm, the landlord had provided a new farmhouse and a complete new array of buildings. The tenant had rearranged the fields, throwing two together here, straightening a fence there and building good service roads. Practically the whole area had been under-drained at ten-yard intervals, the landowner providing the half-million tiles required and the tenant bearing most of the labour costs.

Here again the arable was run, with the least possible deviation, on the four-course system. The regular staff, consisting of thirteen men, was paid with £8 11s. a week, and the total annual wage bill, including casual work on the root crops and for the hay and corn harvests, was about £580. The tenant was especially complimented on the completeness and accuracy of his accounts, which provided an actual basis for a judgment of the management " with a view to profit." Needless to say, such a basis was rarely available.

The winner of the second prize was specially commended on account of the high wages which he found himself able to pay. The regular men workers had 12s. a week in winter and 16s. 6d. in summer, while the youngest lad thought himself very rich on his sixpence a day. The results seem to have justified the expense, for the farmer had a contented, loyal and industrious staff at a time when discontent was widespread. Joseph Arch was already laying plans for his Union of Labourers and its formation, in the following year, was the prelude to a period of agitation and labour disputes.

Cheese was still the main product of the dairy farms of the area, and the skill as a cheese-maker of the farmer's wife was one of the bases of the awards. The winning farm, near Oswestry, carried a herd of 50 cows which, in July, were found to be producing an average daily yield of 31½ lb. of milk. There was also a quota of young stock and a flock of sheep, while a considerable amount of land grew corn. Here again there had been notable improvements in the immediate past, more especially with respect to the pastures. There had been draining and liming upon a large scale, but the most striking results had been got by mass applications of bones. It seems to have been ordinary good practice in the district to give half a ton an acre at intervals of about eight years, and, where it was a case of reclaiming really poor grass, a ton or even two tons an acre might be applied.

Lest we should suppose that our grandfathers were all past masters of the act of husbandry we may turn to the report for 1872. This was written by Thomas Bowstead, from highly progressive Cumberland, and applied to the rather backward counties of Glamorgan and Monmouth. There was, indeed, some excellent farming to be seen in those parts, but Mr. Bowstead regretted to have to say that " not a few of the entries were *totally unfit* for an inspection of this kind. Far be it from us to under-value the frugality, industry and sterling honesty practised by some of the less eligible of the competitors ; but surely lands dirty, imperfectly tilled and out of condition ; fences crooked, broken down and three times too wide ; gateways without gates ; buildings low, dark and dilapidated, badly ventilated and inadequate to the requirements of the farm ; live stock ill bred, ill fed and ill looking ; surely these are not the marks of prize farming."

The fact was of course that, even in the 'sixties and 'seventies, not all districts provided scope for farming enterprise, and, therefore, not all farmers had profits to reinvest in improvements. It was for the most part the big men in the arable counties, like Lincolnshire and East Lothian, who, in those days, made Britain's reputation as the best-farmed country in the world. The west and south-west, remote from large markets and " more apt for grass than grain," were relatively backward. In 1872, indeed, the industries of South Wales were expanding fast, but the immediate effect had been to deplete the countryside of labour, and to make the farmer's disability, for the time being, still greater.

The weakest side of the farming of South Wales was the breeding and management of cattle. There were some few good herds of Herefords and, besides these, some of the old Glamorgan and Pembroke breeds that were not to be despised ; but " on the majority of farms nothing but ill-bred mongrels." By contrast there had been notable progress in sheep-breeding, and there were some fine and thriving flocks both of Cotswolds and Oxfords. The efficiency of labour was low, a state of affairs that the judges attributed to excessive consumption of cider ; and since cider allowances ran up to three gallons per man per day, there may have been some foundation for the theory.

In connexion with the Hull Meeting of 1873 separate prizes were offered for the Wold and Holderness districts, but only the latter produced enough entries. Holderness, being a fine wheat country, was prosperous, though handicapped by the difficulty of growing roots and thus of producing manure. There was, too, a lack of good housing both for man and beast.

Bedfordshire in 1874 was labouring under two serious handicaps. There was a severe drought and a general failure of the root crops, and there had been a twelve-weeks' strike of a considerable proportion of the workers. The point at issue was whether the prevailing wage of 13s. a week should be raised by two shillings, as the men claimed, or only by one, which the farmers offered.

A majority of the competitors were tenants under the Duke of Bedford, who had recently spent vast sums in improving and re-equipping his Woburn estate and in building model villages. The first and second prizes went to neighbouring tenants on this estate, while Charles Howard, the famous sheep-breeder and a member of the great firm of implement makers, had to be content with the gold medal which formed the consolation prize.

Most of the competing farms, both on light and heavy land, were still run on ordinary lines, with varying proportions of four-course arable and permanent grass, and carrying herds of Shorthorn dairy cows and flocks of Cotswold or Oxford sheep. The largest enterprise on the winning farm was the making of butter for London market. One of the pioneer vegetable growers, Lester of Kempston, received a commendation, but the judges were at a loss in trying to compare what they regarded as an overgrown market garden with numbers of ordinary farms. Unfortunately they give no description of the new system which was destined to replace the sheep-and-barley farming on most of the light land of the county.

In the Somerset competition of 1875 separate classes were provided for hill farms, for dairies and for others ineligible under either of these heads. The judges were impressed with the quality of the land in the Vale of Taunton and they record that the best of the fattening pastures in the Bridgwater Marshes were letting at rents of £6 an acre or more.

But their general impression was that " as a rule, the general agriculture of the county was backward."

To the modern reader the most interesting of the farms described is that which had the award in the Hill class, one of 200 acres high up on Brendon Hill. The tenant had taken a fourteen-year lease in 1865, at which time all but twenty acres was unenclosed moorland ; in the interval all but a few acres had been enclosed and reclaimed. The process of reclamation began with the paring of the turf by men with breast-ploughs and the burning of the sod ; the ashes were spread, boulders cleared and the land was ploughed and dressed with two tons of lime to the acre. The first crop was oats, followed by swedes or turnips, which had a dressing of bones or superphosphate and were folded with sheep. In the following spring a grass-seed mixture was sown with three or four pounds of rape, which was again folded. The ley was allowed to lie two years, when it was broken up for a repetition of oats, roots and rape. Thus the scheme would have been admirable if only the proper types of grasses and clovers had been available, but the mixture of ryegrass and temporary clovers that was actually used was ill calculated to its purpose. The live stock consisted of a breeding herd of Devon cattle and of a flock of Devon Longwool ewes, the latter being taken to a low-ground farm for lambing.

In 1876 the Show went to Birmingham and the Farm Competition to Warwickshire, but the local farmers were in no great heart for the latter. Joseph Arch was a Warwickshire man and, since the formation of his Union in 1872, his native county had been the storm-centre of the labour troubles. Added to this, the season of 1875 had been very wet— " one of the most difficult of the century." It is not surprising to learn that, when the judges made their preliminary inspection in November 1875, they found many foul fields, and farm work everywhere in arrear. The winner of the first prize seems to have solved his labour problem by giving his men all that they asked and more, his carters having 16s. or 17s. a week with rent-free cottages and various perquisites. At any rate the judges found " a farm perfectly clean, well horsed, well manned, well cropped and well mastered ; difficulties of season overcome ; labourers well paid, well housed, comfortable, tidy and interested in their master's success."

By contrast, the event of 1877, open to Lancashire, Cheshire, Denbigh and Flint, with a separate section for the Isle of Man, was a notable success. With generous local support a prize fund of £350 was raised and there were 46 entries in the eight classes. A Lothian farmer, who might be expected to have high standards, was brought down to help with the judging of the arable classes, and Chalmers Morton, the famous editor of the *Agricultural Gazette*, co-operated with two leading farmers in placing the dairy and stock-breeding holdings. The local enthusiasm was partly explained by the fact that farm competitions were no new thing in the area, especially in Lancashire ; but it was also true that the depression which was beginning to settle on the corn farms of the East and South had so far left the North untouched. The trade of Liverpool and the cotton industry of Lancashire were both steadily expanding ; arable farmers near the towns had a growing market for potatoes, hay and straw, while grassland farmers had corresponding opportunities of selling milk and dairy products, pig-meat and eggs. The custom of tenancy had already been adapted to the changed conditions, most farmers having the privilege of cropping as they chose and of selling away anything that would fetch a price. The fertility of the land was maintained, in the one case, by means of stable and cowshed manure from the towns and, in the other, by the purchase of large quantities of feeding stuffs as well as of bone manures. Wages were high—a guinea a week seems to have been standard in Lancashire—but rents, still the largest item in farmers' expenditure, had hardly kept pace with the growing opportunities of income. The picture that we get from the judges' reports is one of good landowners, enterprising tenants and efficient, industrious workers all pulling together and prospering. What most impressed the judges of the pastoral classes was the quality and productivity of the temporary leys. " The climate and soil together lend themselves admirably to the growth of ' artificial ' grasses, and everywhere bone dust . . . possesses an efficiency, especially on grass lands, unknown to the southern and eastern counties. The early establishment and long continuance of sown grasses is probably the most characteristic and influential of all agricultural features of the district."

With the Show at Bristol in 1878 the farm competition embraced Gloucestershire, East Somerset and North Wiltshire. There were again four classes, for large and small arable farms and large and small dairies respectively. The winning farm in the first class was in the Cotswolds, and of the same type as that which had won the first competition at Oxford ; now, however, steam tackle had largely displaced horses. The dairy farms of the Vales were stocked with shorthorn-type cows and the herds were being rapidly improved by the use of pedigree bulls—for the pedigree Shorthorns of the south were still of deep-milking strains. Most of the milk was, of course, still used for making cheese,—Cheddar in Somerset and Single or Double Gloucester elsewhere.

In 1879, the year of Kilburn, the competition was of a special character. During the preceding decade, or more, many towns had installed modern systems of sanitation, and the problem of sewage disposal was exercising the minds of town and city councils. The Mansion House Committee, accordingly, offered two prizes, each of £100, for the best-run sewage farms dealing respectively with the waste from large and small towns. Of course no form of irrigation was likely to produce impressive results in a year of incessant rain but, even making due allowance for the season, the judges were not greatly impressed with the results being obtained. Sewage was evidently of less value as a manure than they had expected. The subsequent sixty years have gone to show that their opinion was well founded, and that the economic utilisation of modern sewage is a most difficult problem.

Like the judges of 1877, those who made the awards in Cumberland and Westmorland in 1880 were struck by the prosperity of the north country. "I do not believe," says the reporter, "that a district in England could be selected where farmers and landlords have been so little hit by the prevailing agricultural depression of the last few years." Here it was impossible to point to expanding local markets. Rather it was simply that the farmers had always been stockmen, and had shown no reluctance to get out of the unprofitable minor business of wheat production. They had restricted their cropping more and more to oats, swedes and grass leys, and had bent their energies to the production of more keep and to the improvement of their flocks and herds. The leys were strikingly successful, partly because of the favourable climate but partly also because the seeds-mixture represented a great advance on those commonly employed elsewhere ; cocksfoot, timothy, meadow fescue and other 'natural' grasses were in general use wherever the sward was intended to lie two years or more.

Another point that struck the judges was one that is often mentioned by nineteenth-century writers on English farming ; this was the contrast, both in bodily strength and vigour and in mental alertness and ambition, between the farm workers of the north and those of the south ; it had, in fact, become a trite observation that the north-country labourer was the better value to his master at his substantially higher wage. Most writers seem to have thought that the northerners were the better breed of men, but it is to be noted that the contrast is not drawn by writers of earlier times. Little doubt now remains that the explanation is to be found in the different dietaries of the two regions ; the oatmeal, potatoes and milk which formed the basis in the north country provided a complete and well-balanced ration, whereas, in the south, the increasing dependence on white bread must inevitably have resulted in widespread mineral and vitamin deficiencies.

The picture of Derbyshire farming, as presented by the judges of 1881, is a strange mixture of light and shade. Wheat had been monotonously unprofitable for the past seven or eight years and in the districts less suitable to its cultivation the crop had been almost given up ; oat crops had suffered heavily from an insect pest—probably there had been a particularly large ' wave ' of frit-fly ; on the other hand there had been a good and sustained demand for barley by the Burton maltsters and, with some good harvests in the Midlands, many samples had sold at good prices. The wet seasons had brought rushes into many pastures and losses from liver fluke in sheep had been so heavy, especially in 1878–9, that flocks were a good way below normal numbers. Many farmers, however, had found salvation in the expanding demand for milk ; on one farm after another the cheese-room was filled with lumber and the utensils covered with dust. " The Midland Railway informed us that in 1872 they carried over their whole system about one million gallons of milk and that this year (1881) they will convey five and a half million gallons." With the more profitable outlet for their product, dairy farmers were intensifying their methods of production—increasing their dairy herds and buying much larger quantities of feeding stuffs, especially brewery grains and decorticated cotton cake. One of the competing farms could show an average yearly yield, from 36 cows, of 921 gallons. The chief complaint of the dairy farmers was that pedigree breeders of Shorthorns had ceased to provide for the needs of the home dairy farmer. " Now and then we came across a pedigree bull, but we were continually told that those dairy farmers who relied upon the herd-books for their bulls ' soon pedigreed their milk away.' "

The judges' report on Berkshire for 1882 contains little in the way of general comment, but one remark on farm accounts is worth quoting : " In some instances sound practical farmers had for the first time attempted book-keeping in connexion with this competition and by this time have probably abandoned it as a hopeless task ! "

All Yorkshire was included in the area for 1883, and the county was evidently having a mixed experience. Much of the heavier arable land was weedy and out of condition, but Wold farms were still, for the most part, well tilled. Prices for meat had been fairly well maintained, but the county had had a specially unfortunate experience with foot-and-mouth disease, and those farmers who had had two or three successive visitations at short intervals were in no happy case. The live-stock farming seems to have been much as it now is—except, of course, that more beef and less milk were produced in those days. We read, for instance, of Leicester ewe flocks, crossed with Down rams, on the Wolds ; of Lonks and other Blackfaces in the higher Dales and of the crossing of these with the Wensleydale ; and of the importation of draft half-bred ewes ("Bamshires") from the Scottish Borders.

There was some difficulty in securing local support for 1884, but eventually the competition (for Shropshire, Hereford and Staffordshire) was quite successful, with 21 entries in the three classes. Here, as elsewhere, it was a time of "up horn and down corn." Land was being laid away to grass, and much of this has not since seen the plough. Whereever milk could be sold, dairy herds were growing fast ; elsewhere farmers were breeding more Hereford cattle and Shropshire sheep and, by more intensive feeding and the breeding of earlier-maturing strains, were still further increasing their output of meat.

Interesting evidence of the progress of stock farming was discovered in the records of the valuations of a particular farm in 1822 and 1884 respectively ; the figures were :

	1822		1884	
	Number	Value	Number	Value
Cattle	45	£140	72	£1,332
Horses	13	£132	19	£510
Sheep	104	£156	277	£842
Total values . . .	—	£428	—	£2,684

A new feature of the competition, introduced in this year, was the award of medals and certificates to farm workers of exceptional skill and industry employed on the competing farms. Six such awards were made.

With the Show at Preston in 1885, Lancashire, Cheshire and North Wales had their second turn. The judges could still congratulate the district on its prosperity. Wheat, which had used to be thought cheap at any figure below 50s., had fallen to 33s., and the prophets predicted that worse was still to come. But so long as the straw could be sold at £3 or £4 a ton the bulky crops of Lancashire and Cheshire brought in a fair total income. Moreover, it had been possible to expand the potato acreage without glutting the market, and a simple three-course of wheat or oats—hay—potatoes produced a large gross return. At any rate the standard of farming was being maintained at a high level.

Norfolk and Suffolk, as seen in 1886, were showing signs of the times, and reductions in rents had been fairly general ; yet we do not get an impression of extreme distress, at any rate in the light-land areas. The choice barleys which were produced brought good money, and wherever root crops could be grown with certainty there was still something to be made out of the winter fattening of cattle and sheep. Norfolk no longer occupied its old place as the home of all that was best in farm practice, but there was still much to be admired. The judges thought that, in places, the one-year clover crop might perhaps with advantage give place to a two-year ley, but they found nobody turning to this expedient.

Reversion to the long ley, rather than the conversion of arable to grass, was in rapid progress in Northumberland and Durham in 1887. The high corn prices of the 'sixties had tempted many of the low-ground farmers to plough more ground, and four-course farming had become common. Now the three-year ley was coming back, though there was still a good deal to be learnt about this system. Some farmers, indeed, were making use of cocksfoot and timothy, but, even so, the sward deteriorated very seriously in the third year and failed either to produce its proper quota of pasturage or, when it was ploughed under, to add much fertility to the soil. The magic that lacked was, of course, wild white clover.

The judges spoke very highly of the sheep stocks, both the hill flocks of Cheviot and Blackface and the Border Leicester by Cheviot crosses (half and three-parts bred) which were universal on the low-ground and semi-upland farms. On the other hand, there was little cattle-breeding, and the Irish stores upon which most farmers depended are described as of very poor quality. If Ireland could provide no better raw material, the district, the judges thought, must produce its own.

The competition of 1888 covered Lincoln and Nottingham, both of which counties had things of great interest to show. The former was now generally spoken of as the premier agricultural county of England ; farming on the Heath and Wold had wonderfully maintained its high standards, the fatting pastures along the coast were still letting at high figures, and the cropping of the silt land, with the potato acreage steadily expanding, was beginning to take on something of its modern character.

The 'Lincolnshire Custom' had long been regarded as the best and most equitable in the country, and had formed the basis of the earlier Agricultural Holdings Acts. The local breeds of live stock—the Lincoln Red, the Curly-coated pig and the Lincoln Longwool sheep—were all greatly admired although, with regard to the last, there was some doubt whether the quality of its mutton would continue to satisfy the increasingly fastidious taste of the consumer. Already there was some talk of a demand for smaller joints. The pigs were mainly used to produce the 'mountainous' carcases of fat bacon that provided one of the almost universal perquisites of the farm worker.

In Nottingham, rents had fallen by 25 or 30 per cent. in the preceding ten years and farmers had changed the emphasis of their farming from corn to meat. There was little good pasture, so that the chief business was winter fattening.

Most interesting is the account of the farming on the poor sands which had been reclaimed, in the earlier half of the century, from the waste of Sherwood Forest. Remarkably enough, it was two such farms, and not those on the better lands of Lincolnshire, that won the awards in the large arable class. The system on this unpromising soil was simple enough. The land was dressed, at ten-year intervals, with three or four tons of lime to the acre; "without this the land would probably be worthless." The root crops, chiefly swedes, had dressings of dung and phosphates and, latterly, of kainit as well; then followed barley with grass-seeds, the ley being left for two years and ploughed up for wheat or oats. Swedes were a reliable crop—the winning farmer had a 30-ton yield in the year of the competition—and were used, along with heavy rations of cake, to fatten yarded bullocks and folded tegs. The system, of course, stood or fell according to the profits or losses from winter feeding, for the productivity of the corn crops depended on the manure from a large head of stock. But in 1887–88 both cattle and sheep had paid well, and farmers were in cheerful mood. The winning farm, which had been a sandy waste forty years before, had in that year turned £2,600 worth of store sheep and cattle into £4,900 worth of meat.

There was no competition in the year of the Jubilee but the series was continued in 1890, the area being Devon and Cornwall. The authors who wrote on these counties in the earlier volumes of the *Journal* (in 1845 and 1848) had found them very backward, and had condemned many of the standard practices as 'slovenly' and 'barbarous.' But much had happened in the interval, and in 1890 there was evidence of progress and enterprise. Naturally, the development of the railway system had meant much to this far corner of the country ; on the one hand, it had made possible the marketing, in London and the Midlands, of such produce as broccoli, early potatoes and strawberries, and, on the other, it had turned many of the coastal villages into large holiday resorts which provided a summer market for milk, meat and vegetables.

The judges still found some practices that they could only condemn. The seeds mixtures used to produce the all-important temporary pastures were often ill-selected, and the ingredients often of poor quality and even adulterated. Many orchards, both of cider and other fruit, were completely neglected. Some of the rotations were ill-calculated to keep the land clean, and consequently weeds, especially thistles, were too abundant. On the other hand, there was nothing but praise for the local breeds of stock, whether of sheep or of cattle, and some of the competing farms were examples of all-round good husbandry.

The second Yorkshire competition of 1881 produced the disappointing total of only eleven entries in the three classes. English farming was now feeling the full weight of depression and, for the first time in the competition reports, we read of tenantless farms. These were specially common in the district round Doncaster, where the Show was held. The particular difficulty here was that local custom meant an exceptionally heavy in-going valuation for new tenants. Other evidence of the difficult times is quoted in the records of the money realised for corn from a farm in the Wolds. In 1877 the figure had been £2,700 and, with a bumper harvest, it still reached £2,500 in 1884. In 1886 it was £1,030, and two years later, with a poor crop and the price of wheat down to 31s., it was no more than £895. Farmers who had " seen the red light " and had turned in time to grass and stock were in the least bad position ; those who had struggled on with their traditional systems had lost the capital which reorganisation would have required. The most successful farmer among the competitors, from the financial point of view, was a breeder of pedigree Shorthorns and Leicester sheep.

The first series of competitions was brought to a close with that in Warwickshire in 1892, and produced the very satisfactory entry of 26. Grain prices were better in 1891—wheat rose to 37s.—but both beef and mutton had fallen, and it was now clear that serious overseas competition had to be faced in the meat as well as in the corn market. The late harvest and wet autumn of 1891, followed by a winter that killed beans and oats and, in turn, by a spring drought, added to the anxieties of the year. The surprising fact is that the judges were able to report that there was still some excellent farming to be seen.

FARM PRIZE COMPETITION

In 1892 the Journal Committee recommended that the scheme be suspended for a time ; the competitions had now covered most parts of the country and had served their immediate purpose ; moreover, the prevailing conditions were not conducive to prize farming. They thought it would still be useful to publish, year by year, some account of the best practice in the Show area, but they proposed to appoint a " commissioner " to visit and describe typical holdings. There was opposition to these proposals but the Council, by a large majority, approved them. It was not until the Park Royal situation had been cleared up that the farm prize scheme was revived.

The first of the new series was held in Lincolnshire in 1907. In the interval farmers had, one way or another, found some sort of solution to their economic problems, and for the last five years the industry had been almost prosperous. The Lincolnshire competition, with generous financial support from Sir Richard Cooper and from the local Hunts, was the largest and probably the most successful ever held. There were 105 entries and " Interest was great and competition keen. Undoubtedly the farming of the county has benefited ; many neglected buildings have been put in order, gates renewed or repaired, fences put straight, ditches cleaned. . . . The land has been better cleaned, better tilled and many improvements made."

The first prize in the class for large holdings went to Mr. John Evens for his farm at Burton, the tenancy of which had passed from father to son over a period of nearly 200 years. The judges were full of praise for the crops and pastures and for the flock of Lincoln sheep. But the outstanding feature was the herd of Lincoln Red cattle which had been fully milk-recorded since 1885. " By keeping milk records, by weeding out indifferent milkers and by using bulls from dams of proved dairy merit, the owner has succeeded in building up what many visitors have described as the best herd of dairy cattle they ever saw." Thirty-two years later Lincolnshire was to rejoice to see the winner's pioneer work recognised by his election to the Presidency of the Society.

Heath, Wold and Fen were all represented in the prize list. The judges' tours were all done by motor-car, at a vast saving of time, and with no more serious mishap than two punctures in their two thousand miles of journeying.

Northumberland and Durham in 1908 produced a fair competition, though the exclusion of holdings larger than 600 acres prevented the judges from seeing some of the finest farms in the former county. Durham took all three first prizes, though Northumberland had the seconds. The winning large holding was one of the intensively-run dairy and arable holdings on the good land near Sunderland ; in the second class the winner was Mr. George Harrison of Gainford Hall, Darlington, whose fine head of Shorthorns, old-established flock of Leicester sheep, heavy crops and well-managed pastures combined to make an impressive whole.

In 1909 there were two distinct competitions, the one for Gloucester and Wiltshire and the other for Hereford and Worcester, with four classes for each area and a total prize fund of £660. The total entry was seventy-five, despite the failure of the class provided for small market-garden holdings.

Perhaps the most interesting of the many good entries was that which won the hundred pounds for the best large holding in the Gloucester-Wiltshire section. This was a chalk farm of 750 acres near Salisbury, farmed on almost precisely the system that has been so entertainingly described by Mr. A. G. Street in *Farmers' Glory*. There was the same large flock of Hampshire sheep, with the same intricate cropping scheme to provide its daily fold ; the dairy herd, which the farmer regarded as a necessary evil ; and there was a block of water meadow under its " drowner."

Land was still going down to permanent grass, and for this purpose farmers had turned to seeds mixtures of the most extreme complexity. One, which is given in full, contained nineteen separate species—while still omitting the wild white clover which, as was still not generally realised, would have been worth as much as a dozen of the sorts included.

The famous farm of Maisey Hampton, where was one of the most noted of the few herds of recorded Dairy Shorthorns, could win no more than a commendation.

In the report on the Hereford and Worcester section there is an interesting account of hop-growing in the Teme Valley, though with a rather gloomy reference to the decline of the industry.

Lancashire and Cheshire had their third competition in 1910. Suburban arable farming had become still more intensive since the last visit, many farmers having a third or more of their land under potatoes, and the early crops being now commonly intercropped with cauliflower or cabbage. Already, however, there were complaints about the scarcity of the town manure upon which the system had so long depended, and a good many farmers were being obliged to go back into live stock. Cheshire dairy farming had also been intensified ; many farmers had given up cattle-rearing in order to make room for more milking stock, and heavier purchases of feeding stuffs were the rule. The first-prize farm, near

Nantwich, extending to 209 acres, had raised its cow stock from 75 to 95 head in the preceding three years. In 1909, the milk sales, at about 7½d. a gallon, had brought in £2,475.

The general scheme of things in Norfolk and Suffolk had changed but little between 1886, when the first farm competition had been held, and 1911, when the Society returned. Some 5 per cent. of the arable had gone down to grass, and bits of the Breckland had been abandoned altogether ; sheep flocks had fallen by some 10 per cent., and wheat, barley and root crops by a little more. On the other hand, there had been some expansion in the acreages of cash crops such as potatoes and picking peas, and in both counties there had been considerable increases in the numbers of grazing cattle and of pigs. The old meticulous care had hardly been maintained—there were untilled headlands and twitchy corners that would, in earlier times, have been thought disgraceful ; but things were on the mend. The Suffolk ewes that constituted the bulk of the breeding flocks were commonly mated to Cotswold or Lincoln rams. The big Irish bullocks that filled the yards were still fed on the traditional ration of a hundredweight or more of roots and up to a stone of linseed and cotton cake, along with hay. Wages had remained at a relatively low level, daymen getting only 13s. a week except in harvest, and skilled stockmen about 17s.

The 65 entries were classified according to size, and all classes were well filled. Small holdings numbered 9—easily the best entry ever produced. The winning large farm was at Playford, near Ipswich, already well known for its Suffolk flock and beginning to earn fame for its Dairy Shorthorns. The prize-winning small holdings were found in the Wisbech district, all being devoted to fruit or to fruit and vegetable production.

The Yorkshire competition of 1912 produced 49 entries. No striking innovations in farming were seen, but general efficiency was improving, partly at least as a result of the good work that was being done, in the way of assistance and advice, by the Yorkshire College at Leeds. The Dales, the Vale of York and the Wolds were all represented in the award list. In the arable districts large numbers of Irish cattle were still fed in winter and the reputation of these was now high ; the best Irish were as good as any feeder could wish to have.

In 1913 there was again a double event, one set of classes for Gloucestershire and the other for Somerset and Dorset. In the report is an interesting review of the change that had occurred in the thirty-five years since the previous Bristol meeting. The arable acreage had fallen by about a third and wheat production by more than half, while dairy cows had increased by nearly a quarter. Coincidently there had been a great fall in employment, reaching 33 per cent. in Somerset. Even the chalk and limestone farms, though they still maintained their flocks of Hampshire or Oxford sheep, were turning to milk wherever an area of reasonably good pasture could be made; but the dairy herd management of some of the newer milk producers left a good deal to be desired. Milk recording had begun, but had not made much progress, and " the vaguest notions prevailed as to the nature and quantities of the rations fed." Some good dairy herds were being " topped " with beef Shorthorn bulls with, naturally, rather disastrous results.

In 1914 Shropshire was the centre of the contest for the third time, being joined on this occasion with Staffordshire and Montgomeryshire. Many well-known names appear in the award list—William Everall with his Shropshires, Morris Belcher with his Shires and William Nunnerley, another breeder of pedigree Shropshires, who, incidentally, had won his class thirty-five years before. The general impression of the judges was of most excellent flocks and herds and of skilled arable farming, combined, however, with some rather indifferent grassland management. With few exceptions the farm accounts left a good deal to the imagination.

The last of the series, held in connexion with the war-time show at Nottingham, was open to the counties of Nottingham, Leicester and Derby and, for the time, produced the satisfactory entry of 21. As on an earlier occasion, the prize-winners in the arable classes were mostly found on the Nottingham sands, but the grazing class was led, as might have been expected, by one of the famous bullock-feeding farms in the Market Harborough district of Leicestershire. This was Tur Langton Manor, where 450 cattle, besides some sheep and some notable Shire horses, all found abundant food upon 400 acres.

The competition would have been continued but for the War, and there was some disappointment when it was decided to omit it in 1916. But war conditions were hardly conducive to ' Show Farming.'

The scheme undoubtedly was worth what it cost, and its revival at some future time might well be considered. The reports are of the greatest historical interest—illustrating, as they do so fully, the diversity of this country's farming industry and, above all, the resourcefulness of its farmers in meeting the great changes and difficulties which the period brought.

CHAPTER XII

The Journal

IT will be remembered that the founders, when they originally set out the objects of the Society, put first on their list the dissemination of agricultural knowledge, and it is evident that they had in mind, as the chief means to this end, the publication of a Journal. In fact, until the Show developed, this was perhaps the most important of the Society's enterprises. The *Journal* met one of the greatest needs of the times ; it provided a record of discovery and invention and, as Arthur Young's publications had done in the beginning of the century, made known the best practice of one district to farmers in others.

It was a most fortunate thing that the Society should have found a man of quite outstanding ability and vision to take charge, from the outset, of this part of its affairs. This was Philip Pusey, who, in December 1839, was appointed Chairman of the Journal Committee and who, for the next sixteen years, was to give a great part of his time to its editorial control.

The formation of the Society found Pusey, at the age of forty, well on the way to a distinguished career in politics. He had had the usual upbringing and opportunities of the eldest son of a fairly wealthy land-owner of the time—had been to Eton and Oxford and had done a long tour of the Continent—and he had developed a wide range of interests, both in scholarship and in practical affairs—" He wrote on philosophy for the *Quarterly Review,* on current topics for the *Morning Chronicle* and on farming for the Royal Agricultural Society." He was the close friend of Gladstone, of Thomas Carlyle and of Sir Robert Peel. Baron Bunsen, who spent a great part of his stay in England as Pusey's guest, described him as being " a unique union of the practical Englishman and the intellectual German."

Pusey represented Berkshire in Parliament from 1835 till 1852 and was, as Disraeli once said in debate, " one of the most distinguished country gentlemen who ever sat in the House of Commons." It is true that his greatest effort as a politician ended in failure, but this was because the idea which he strove to get embodied in English law was produced twenty years before its time. This idea was that which underlies the modern law of agricultural tenancy—in Pusey's own words, " that a tenant-at-will should have security for the capital which he ought to be encouraged to invest in the soil." He laboured hard, from 1847 till 1852, to persuade Parliament to his opinion, and eventually the Commons passed a bill which went part of the way ; but this was thrown out by the House of Lords and nothing effective was done for another twenty years.

On his own estates in Berkshire Pusey practised what he preached, adopting in his farm agreements the Lincolnshire custom of which he thought so well, and which was to form the basis of the Agricultural Holdings Acts. In other respects, too, he set an example in wise and liberal land-ownership, encouraging his tenants to improve their methods and taking the closest interest in the comfort and well-being of every worker on his estate.

Like Spencer, Pusey sacrificed many personal ambitions in order that he might work for the advancement of agriculture, and he gave the Society's affairs first claim upon his time. He was its President in 1841 and again in 1853, but it was always the *Journal* that was his special interest and concern. Sir James Caird said of him that he was " the leading agricultural writer of his day," and his own contributions, which make a long list, set a very high standard ; his enthusiasm shone through everything that he wrote. In a very few years the *Journal* established a high reputation and in 1844 the *Quarterly Review* said of it :

The *Journal* of the Royal Agricultural Society will be a permanent monument to the honourable member for the County of Berks, whose patriotic earnestness of purpose induced him to take on himself the gratuitous labour of its editorship. The same spirit of zealous endeavour to assist in teaching the farmers of England to meet the necessities of the times has prompted Mr. Pusey not only to prepare for the press the contributions of others but also to enrich its pages with several most instructive articles from his own skilful pen.

Pusey was not only an excellent writer but was a very capable judge of the value of the work of others, and he was assiduous in his search for good material. A large proportion of this was obtained in the form of the Prize Essays, some of which were by professional writers, while the majority were the work of farmers, land agents and landowners. These essays, of course, covered practically the whole field of farming, and their general literary standard was so high that it is clear that Pusey must have spent laborious days in dressing up his wares. The most interesting of the Essays, to the general reader of to-day, are those which reviewed, county by county, the recent progress and the existing state of English farming. This series covered more than thirty counties during Pusey's time as Editor, and it was concluded only five years after his death. It presented the only detailed picture of our farming that had been drawn since the County Surveys of the old Board of Agriculture, published in the years about 1810.

Apart from Prize Essays, the early volumes of the *Journal* contain many of the scientific lectures delivered to general meetings of the Society, and various articles by specialists. The longest series of the latter was that by Curtis, describing in detail the important insect pests of all the major crops. This series formed the first complete treatise of its kind.

Lastly, apart from the reports of the Society's own activities, Pusey's *Journal* included large numbers of short practical articles of the kind that would now find their place in the weekly agricultural press. In particular, Pusey was keenly interested in practical field experiments, and was not only prepared to give publicity to their results but often made suggestions about trials that the farmer might carry out.

Pusey not only induced the old and experienced farmers to impart some of their wisdom, but also encouraged young men of talent to try their hands at writing. Among these were three who later became the leading agricultural journalists of their time and who, among their other activities, were to do a great deal to help Pusey's successors. These were John Algernon Clarke, Chalmers Morton and John Coleman.

Clarke, the son of a fenland farmer, began his career as a writer at the age of nineteen when, in 1847, he won the Society's £50 prize for an essay on " The Great Level of the Fens." Three years later he was again successful with a paper on " The Agriculture of Lincolnshire." In 1857 he became agricultural reporter for *The Times*, continuing, however, to undertake frequent articles and reports for the Society. Ultimately he took over the editorship of the world's oldest agricultural newspaper, *Bell's Weekly Messenger*.

Coleman was the first pupil of the Royal Agricultural College. His first appearance in the *Journal* was in 1855, with a Prize Essay on Soil Fertility. In the following year he returned to his old College as Professor of Agriculture, and from that time onwards did an increasing amount of work for the Society—as judge of implements and of Farm Prize Competitions, and as a reporter on implement trials and on the development of steam cultivation. He later became agricultural editor of *The Field*, and was the author of a valuable series of books on British live stock.

The greatest of the trio was Chalmers Morton, the son of Lord Ducie's agent in Gloucestershire. He studied under Professor Low at Edinburgh, and afterwards for some few years was a farm manager under his father. His first contribution to the *Journal* was a letter to Pusey on the cultivation of carrots, which appeared in the second volume. In 1843, at the age of twenty-two, he was chosen to start and direct a new agricultural weekly, and in a few years he made his *Agricultural Gazette* the leading organ of its kind. Its prestige was fully maintained until his death. His Encyclopædia of Agriculture was another monument to his industry.

Among the non-professional writers were several members of the Council, including H. S. Thompson, who was later to fill Pusey's shoes.

Since the *Journal*, in its early days, had to perform some of the functions of a newspaper, Pusey was in favour of publication at short intervals, and seems to have hoped to run a quarterly. In fact he had to be content, from the first, with three issues a year, and even this arrangement meant too heavy a cost of distribution. After three years the *Journal* was reduced to two parts, issued in spring and autumn, and this arrangement remained in force until the Jubilee.

In 1853 Pusey was given the honorary degree of D.C.L. by his old University and, about this same time, his name was considered for a peerage. But now his health was giving way under the strain of incessant restless work. His last years were darkened by the hopeless illness of his wife, and he had to remain by her bedside during the Country Meeting of his second year of the Presidency. A few months later Mrs. Pusey died, and in the following year Pusey himself suffered a paralytic stroke. He died in 1855 at the age of 56. It can truly be said of him that he gave his life for the cause of agriculture.

Pusey's place as Chairman of the Journal Committee was taken by H. S. Thompson, whose work in connexion with the implement trials, and whose incursion into scientific research, have been mentioned in earlier chapters. He and Brandreth Gibbs now became the two most active members of Council, the former as organiser of the Show and the latter in the other side of the Society's work. Thompson was neither as great a scholar nor as good a writer as Pusey, yet he had other qualities that Pusey lacked. In particular he had a keener eye for inventions and discoveries, and he was a sounder judge of their likely practical value. For instance, he followed the Rothamsted experiments, year by year, with the closest interest, and was fully convinced of the value of the work at a time when many farmers were still sceptical of any useful outcome. It was doubtless through his friendship with Lawes and Gilbert, and his understanding of their objects, that so many of their classical papers found a place in the *Journal*.

Two other members of Council were appointed to help Thompson. The one was Dyke Acland and the other that wittiest of all agricultural journalists, Chandos Wren Hoskyns, whose articles in the *Agricultural Gazette* (collected as *Talpa, or the Chronicles of a Clay Farm*) we may still read with great entertainment. In 1859, however, Thompson was elected to Parliament, and neither of his understudies could take over the duties that he had performed. It was therefore necessary to appoint a paid editor. This was P. H. Frere, who, from this time onwards till his death in 1868, wrote most of the official reports and looked after the details of production.

Thompson, however, continued to take a very close interest in the *Journal*, and to contribute to its pages, until 1873, when a breakdown in health obliged him to give up all his public work. The most important of his own papers appeared respectively in 1858 and 1864. The former is the first of his efforts to set out the principles of grassland management, a subject in which he was specially interested and which, up till that time, had received but little attention. One of the views that he put forward—that mass doses of phosphates are of more use than " little and often " applications—was to get its experimental proof only half a century later. The second paper is a very competent review of the progress of English farming during the first twenty-five years of the Society's life. Perhaps the least successful of his adventures in authorship was his paper on Potato Blight, published by Pusey in 1846—i.e. on the first appearance of the disease. In this article Thompson advanced the theory that the blight was only another manifestation of the disease that caused tubers to rot in the pit and also produced the trouble which was then called " The Curl," but which, as we now know, included a whole collection of virus infections. Even here, however, there is a curious grain of truth in Thompson's bushel of error, for it is now becoming clear that susceptibility to blight is increased by certain forms of virus infection ; this explains why a new variety, found at first to be highly resistant, has, so often, become quite susceptible to Blight after a few years.

In 1866 the Journal Committee decided to abandon the Prize Essay system, having found that, upon the whole, they could get better articles by direct approach to men of known ability and knowledge. The last of the prizes was awarded to that most interesting character H. H. Dixon (" The Druid ") for an essay on the Leicester Sheep.

Philip Pusey

Sir Harry Meysey Thompson

H. M. Jenkins

Sir John Thorold

In 1868, when Frere died, the Council considered the appointment of a successor in relation to the general question of the Society's official staff and decided, after long consideration and full debate, to combine the offices of Secretary and Editor. This decision involved the termination of Hall Dare's engagement as Secretary, and this step was taken with regret, for apparently there was no question of his efficiency in the administration of the Society's affairs. But any doubt about the wisdom of the change was set at rest by the striking success of H. M. Jenkins, who was appointed to the double post.

Jenkins was only twenty-six when he came to the 'Royal.' He had been noted by his schoolmasters as an extraordinarily brilliant boy, but it seemed that he had no chance of a University education, and he was taken away from school, at the age of fourteen, to start work in a corn-merchant's business. This employment he had to give up on account of an asthmatical complaint from which he suffered all his life. He next found work with a firm of manufacturing chemists and here he developed a keen interest in science. By a happy chance he was offered a small post as assistant in the museum of the Geological Society and he seized upon the opportunity of getting to London and seeking, in his spare time, the scientific education that was his ambition. His ability and enthusiasm soon attracted notice, and he rose to be assistant secretary of the Geological Society and editor of its *Journal*. In his leisure time he not only picked up a wide range of scientific knowledge but acquired a mastery of several modern languages.

Jenkins thus had the combination of administrative and editorial experience that the 'Royal' Council required ; but he professed to no knowledge of agriculture, and it was doubted whether he could adapt himself to his new environment. Chalmers Morton, who now had more influence on agricultural opinion than any other man in England, thought that the appointment was a bad one.

But Jenkins had a rare combination of qualities—a scientific mind, a good literary sense and a marked capacity for business administration —and these gifts enabled him very quickly to rise to the occasion. He set out at once to fill the gaps in his knowledge and, by touring the countryside at every available opportunity, he got to know the land of England, and its farming community, better than many who have been through a long apprenticeship in agriculture. One of the chief means towards his education was the series of tours that he made in company with the judges of the Farm Prize Competitions, who found in him a most apt pupil. Later he travelled widely on the Continent and became, as has been said earlier, the leading authority on overseas agriculture as well as on agricultural education.

As far as the routine administration was concerned, Jenkins took the line of delegating a large share of his responsibility to his subordinates, with whom he maintained the happiest of relations. He could go to Denmark or Germany, or could be laid aside by illness for weeks on end, with no fear but that his staff would carry out their allotted duties as conscientiously as if he had been standing over them. As for the *Journal*, it reached, in his hands and under the guidance of Thompson and his successor, probably its high-water mark.

Thompson was followed, in 1874, by a man of his own choice and of his own training who also, by an odd coincidence, filled a second of his master's positions—that of Chairman of the North Eastern Railway. This was a fellow Yorkshireman, J. D. Dent, who had joined the Council at the November meeting of 1861, the first and last over which the Prince Consort presided. It was soon apparent that Dent was to be an acquisition to the Council, and more especially to its committees, where his clear mind, his patience and his capacity for sensing the general opinion of his colleagues showed to great advantage. He was an active member of many committees, but his main interest was in the scientific and educational side of the Society's work rather than in its Show. He wrote less for the *Journal* than either of his predecessors ; his most considerable article, on the " Condition of the Agricultural Labourer," appeared in 1871, while he was still acting as Thompson's understudy.

An exceptional undertaking by the Journal Committee was the publication, in 1878, of a *Memoir on the Agriculture of England and Wales*, prepared for the International Agricultural Congress in Paris. This was included as part of the *Journal*, and was also issued separately as a book of over 600 pages. The first section is a general account of British agriculture by Sir James Caird, which was later republished by him in an expanded form and which, especially if read in conjunction with the same author's *British Agriculture in 1850 and 1851*, records the remarkable progress that had been made in a single generation. J. A. Clarke wrote on Practical Agriculture, Chalmers Morton on Dairy Farming, and Voelcker on Agricultural Chemistry, while other authorities covered Taxation, Land Law, Labour, and the more specialised branches of the industry. The last section, by Jenkins, gave an account of the work of the Society. The book is a rich mine of information for the historian.

In 1881 Dent was promoted to the Presidency, and his place at the head of the Journal Committee was taken by Earl Cathcart, who carried on with uniform success his " half-yearly epistle from Hanover Square . . . the bond of union between the members of a great national society." But in the years before the Jubilee, Death dealt very hardly with the ' Royal,' and his hand fell most heavily upon the band of workers who had made the *Journal* what it was. The volume for 1885 contains Dr. Gilbert's memoir of Augustus Voelcker, and that by Chalmers Morton of Sir Brandreth Gibbs. In the next, Chalmers Morton had to take up his pen again to perform the same melancholy duty for Jenkins ; and in the next again we find a new Secretary and Editor writing memoirs of Coleman, Algernon Clarke and Chalmers Morton himself, all of whom had died within a period of a few months. Most of these men, it is true, died full of years, but Jenkins' passing, at the age of forty-six, left a great gap that the Council had made no preparations to fill.

Ernest Clarke, who was selected from more than a hundred applicants for the secretaryship, had spent nine years as a civil servant, and six more as one of the assistant secretaries of the Stock Exchange and as editor of one of its publications. He was a man of undoubted ability and had the same combination of administrative and editorial experience that Jenkins possessed. Unlike Jenkins, however, his interest in agriculture was that of an antiquarian and historian. Moreover, he was essentially an office man, and found neither interest nor pleasure in tramping over farms or looking at bullocks. With the ordinary farmer members of the Society he acquired a reputation for snobbishness, and it is true that he spent a good deal of time in cultivating the acquaintance of men of position and influence. It seems, however, that, in doing so, he was thinking more of the prestige of the Society than of cutting a personal figure in the world. In the work of the office Jenkins' rather free-and-easy methods had to give way to the more rigid rule and system of the Civil Service. Indeed, all that Clarke did, or that was done under him, was done with meticulous care and accuracy ; but the Committees of Council missed Jenkins' encyclopædic knowledge. As for the *Journal* during Clarke's editorship, it maintained its old level as literature, but it rather reflected Clarke's lack of interest in the ' dirty-boot ' farmer, and a lack of appreciation of the desperate struggle that he was having to wage at the time.

The main interest of the volumes that Clarke edited is in the historical articles which they contain. Sir James Caird's *Fifty Years' Progress in British Farming*, written in connexion with the Jubilee and published in 1890, is an interesting retrospect by one who had been a close observer of farming changes during the whole period of which he wrote ; R. E. Prothero (Lord Ernle) contributed a notable article on " Landmarks in

British Farming " ; Earl Cathcart wrote on Jethro Tull ; and Albert Pell tried his hand as a historian with a short biography of Arthur Young. One's chief regret about Pell is that he wrote so little, for he had a gift. Clarke's own articles make a long list. Soon after his appointment he began collecting materials for a history of the first fifty years of the Society and, although the book was never published, some part of the story appeared in the form of *Journal* articles on the foundation of the Society and its early notabilities such as Pusey and Spencer. Clarke held, for three years, the Gilbey Lectureship in the History of Agriculture at Cambridge and published other articles, on the Old Board of Agriculture and its great men, which were evidently based on his Cambridge lectures.

During this time Lawes and Gilbert, either separately or together, continued as frequent contributors, both of reports on the Rothamsted experiments and of more general papers on agricultural science. Lawes also wrote on other subjects, two of his most interesting papers being a review of world cereal production and a paper on the sugar-beet crop. Among other contributors of the time were W. E. Bear, the editor of the *Mark Lane Express* ; Sir Charles Whitehead of Maidstone, who wrote largely on Fruit and Hop Growing and Market Gardening ; and Robert Warrington, a colleague of Lawes and Gilbert and the author of one of the best little books—*The Chemistry of the Farm*—ever written on agricultural science. The rapid progress of veterinary science is recorded in articles by Professor G. T. Brown and Professor McFadyean.

In 1890, with the beginning of the Society's second half-century, the *Journal* began to be published as a quarterly, Pusey's ambition being thus at last realised. The arrangement, however, was to continue for only a decade. The largest volumes in the whole series are those of the early 'nineties, all of which run to upwards of a thousand pages ; the large bulk of the *Journal* meant, of course, that Clarke had to give it a large share of his time, and the very close personal supervision of every detail of the Society's affairs, which it was his habit to give, made him an extremely busy man. In 1892 the General Purposes Committee advised that he be relieved of the editorship and that Dr. William Fream be appointed in his place. Fream was agricultural correspondent of *The Times*, and had already written a good deal for the *Journal*, including reports on the Society's recent Shows. Also he had just completed the textbook on *The Elements of Agriculture* which the Society had commissioned him to write.

Fream's appointment was the complete success that was expected, but unfortunately his period of office was short. It may be that his

qualities. No committee could have had a wiser head, and no editor a kinder or more considerate chief. Five years later C. S. Orwin was succeeded as editor by C. J. B. Macdonald and thus, for the second time, the *Journal* and the agricultural page of *The Times* were in the same hands. Macdonald was the youngest of four brothers who all became prominent in agricultural journalism, and had already a well-established reputation. His interests were more in practical farming and live-stock breeding than in science or economics, and his five volumes reflect these. He died in 1930 and was succeeded by the present writer.

The general plan of the *Journal* remained unchanged until 1938, except for the inclusion in it, from 1933 onwards, of the annual review of agricultural research (*The Farmer's Guide*), which had formerly been published separately. As the Centenary approached the Council decided to review all the Society's activities and to introduce any changes that might seem likely to increase the value of its services to members. As regards the *Journal*, it decided to revert to more frequent publication, issuing three parts annually, of which *The Farmer's Guide* would be one. At the same time the allocation of funds was increased so as to permit of some increase in size, together with better production in a more attractive form.

It will be no easy matter for the Journal Committees of the future to reattain the standards that were set by Pusey and Jenkins, but at least they may hope to be given the financial means to this end, which provision had to be denied to their predecessors of forty years ago.

CHAPTER XIII

SUNDRY ACTIVITIES

AGRICULTURE, in the narrower sense of the word, is closely related to a number of other rural industries and crafts, and the Royal Agricultural Society has, at various times when special efforts seemed to be necessary, taken an active interest in most of these. Again, during the period 1914 till 1918, when many of its ordinary functions were suspended, it played its part in war work. The following brief review covers the more important activities lying outside the Society's main field.

FORESTRY

By contrast with British agriculture, which in the nineteenth century was the most progressive in the world, British Forestry lagged behind that of several continental countries, and for long the position was that England tried to follow while France and Germany led. Among the objects set out in the Society's Charter was " to collect information with regard to the proper management of woods and plantations," but, apart from the publication of some useful articles in the *Journal*, little was done in the earlier years.

It was not till 1881, when the Royal English Arboricultural Society[1] was founded, that any important organised effort was made for the improvement of forestry, and it was not till twenty years later that the two Societies began to co-operate to this end. The first step was to organise a Forestry Exhibit at the second Park Royal Show of 1904, it being felt that the attendance of so many landowners and agents at the ' Royal ' provided a useful opportunity to spread knowledge of scientific forest management. In organising this exhibit the Society's Committee had the assistance of the President of the Arboricultural Society, of the Board of Agriculture in the person of Dr. William Somerville, and of Mr. Daniel Watney of the Surveyors' Institution.

On the first occasion, and also at the two succeeding Shows, the

[1] Now the Royal English Forestry Society.

sole object of the exhibit was to collect and set up a quantity of educational material, and there was nothing either in the way of competitive classes or of working demonstrations. Separate sections dealt with the culture of the more important timber trees ; with propagation and nursery work ; with the management of woodlands ; with pests and diseases ; with the preservation and utilisation of timber ; and with the tools and equipment required in modern forest practice, including especially those used in Germany.

In 1907, at Lincoln, competitive classes were added, both for specimens of various types of timber and for gates, fencing and other articles made by estate workers from home-grown woods. Most of the classes attracted good entries and, as time went on, the classification was extended.

In 1909 the Society, again in co-operation with the Arboricultural Society, took the further step of instituting an annual Woodland Plantations Competition, with the view of encouraging landowners and foresters in the use of improved methods. This Society provided silver and bronze medals, and the sister body gave a gold medal as a champion award. The first competition embraced the counties of Gloucester, Wiltshire, Hereford and Worcester.

The Woodland Competition rapidly grew in popularity, and its scope was widened to include private forest nurseries and also a class for the forest management of whole estates as opposed to that of particular plantations. In 1911, when the competition area included Norfolk, Suffolk and Cambridgeshire, there were fifty entries in all, comprising twenty-two plantations, twenty-one nurseries and seven estates. The classification of the plantations has normally been arranged according to the type of timber (hardwood or conifer) intended to form the final crop and on the age of the trees, though, in some cases where the area has been very diversified, separate consideration has been given to woodlands in upland and lowland situations. The judges' reports of these competitions contain not only detailed criticisms of the particular entries but also, in most cases, broad discussions on forest policy and other matter of great educational value.

The competitions were suspended after 1914 but were revived in 1919, and have since been continued. During the war, as is well known, great inroads were made on this country's reserves of standing timber, and large areas were felled long before their time. The judges of 1919 were surprised to find, in South Wales, " that some timber had been left for future purposes," but it was true that many estates, once prominent in forestry, found themselves with little or nothing to show ; the post-war competitions suffered in consequence. The judges of 1919 consoled themselves with the reflection that many inferior woods, as well as many good ones, had been cleared, and that owners of the former had been given the opportunity of making a fresh start. Unfortunately, as is also well known, the private estate-owner was to find himself faced with greatly increased planting costs, with a fall in farm rents and with a rise in taxation, all of which combined to discourage replanting. Forestry now seems destined to become more and more a State enterprise.

In recent years interest has been added to the Forestry Exhibit by arranging working demonstrations of a variety of crafts, such as hurdle-making, turnery, wood-cleaving and creosoting. A gate-making contest, for estate workers, has also been instituted.

ORCHARDS AND FRUIT

No branch of British agriculture or rural industry has made more remarkable progress, during the past two decades, than fruit growing. In place of the old haphazard methods, which made for a low grade of produce, we have seen the widespread introduction of highly scientific management, and have watched British fruit acquiring a reputation that is second to none. Large areas have been soil-surveyed in order to obviate mistakes in the choice of orchard sites ; remarkable progress has been made in the use of fertilisers ; rootstocks have been classified and the inferior and mixed material has been eliminated ; pest control has been strikingly improved, and modern methods of grading and packing have been introduced. The credit for this revolution belongs largely to the two Fruit Research Stations at East Malling, Kent, and Long Ashton, Bristol.

Soon after these Stations had got to work the Society realised that it would be an important service to spread abroad the new knowledge, and in 1922 the Botanical and Zoological Committee submitted to Council a scheme for a series of Fruit and Orchard Competitions on somewhat similar lines to those for woodland plantations. In that year classes were provided, in Cambridgeshire, for plantations of the more important types both of tree and bush fruits. The competition was repeated in each of the succeeding twelve years, during which period all the important fruit-growing areas had been visited, some of them twice.

The second competition of 1923 was open to Kent, Surrey and Sussex, an area which held a long-established and leading place in the industry, and it is strange to read, now, the rather severe criticism of the general standards that then prevailed. The cultivation of strawberries and black currants was generally indifferent and, as regards apples, little effort was being made to compete with imported supplies. The judges saw no up-to-date packing sheds, and pointed out that the winter market for dessert apples was being left almost entirely to overseas producers.

As we look through the subsequent reports we get evidence of the rapid introduction of new methods and of an all-round improvement in efficiency. In 1929, with reference to black-currant cultivation in Norfolk, the judges wrote :

" We cannot speak too highly of the standard of management which obtains in the plantations that we visited . . . our recommendation to all who think of planting this fruit is to go to Norfolk and study the methods prevailing there."

In 1933, again, when the South-Eastern counties had their second visit, some of the prize-winning plantations were almost beyond criticism.

By this time it began to appear that the competitions had achieved about as much as could be expected of them ; in the latter years there is reference to the growing provision of technical advice and instruction by County Council staffs, to the fact that the work of the Research Stations was becoming widely known and almost as widely applied, and to the notable efforts of the Ministry of Agriculture to raise the standard of grading and methods of packing. Accordingly, the series of competitions was brought to an end with that of 1934.

DAIRY MANUFACTURES

In the Society's early years, and indeed up till the 'seventies, butter- and cheese-making were farmhouse industries. The processes employed were necessarily of a rule-of-thumb kind, for the chemical changes involved were not understood, and the fact that bacteria were concerned in these was quite unknown. Naturally, the quality of the product depended upon the skill of the maker and (in the absence of thermometers and testing apparatus) on the refinement of his or her senses. But even the best makers were apt to meet problems with which they could not deal, and at times turned out very inferior produce.

As with other departments of farming, the Society's first step was to place the subjects of butter-making and cheese-making upon its list of prize essays and to publish the successful dissertations. That on cheese-making is to be found in the *Journal* for 1846, and gives a detailed account of the methods of the best makers in Cheshire. The corresponding essay on butter-making appeared in 1852.

Much of the pioneer work on the chemistry of cheese-making was done by Dr. A. Voelcker during his early years as Consulting Chemist to the Society, and his contributions were published in two long papers in the *Journal* for 1862 and 1863. The first of these set out the principles upon which depended the production of good cheese, and pointed out the commoner mistakes which were made in practice. The second gave the results of Voelcker's own experiments, which dealt with the yield of cheese in relation to the composition of milk and with the methods of manufacture employed, and produced facts and figures on such points as the control of temperature and acidity. His work went far towards placing the whole process upon a scientific basis.

Meanwhile the Society began to provide classes for dairy produce whenever the Show happened to visit a dairying district. The first cheese show was held at the Chester meeting of 1858, and at Leeds, in 1861, provision was made for both products, though the resulting entry was disappointing. From 1875 onwards dairy produce formed a regular feature of the produce section, classes being provided for all the common varieties of cheese and for both fresh and salted butter.

In the *Journals* of the 'sixties and 'seventies there is frequent reference to the progress of dairying, both abroad and at home. The articles of Mr. H. M. Jenkins, the Secretary, in which he surveyed the agriculture of various continental countries, gave accounts of overseas developments, and referred particularly to the development of dairy factories. The first factory in England seems to have been that set up at Derby in 1870 ; in this case an expert was brought from the United States to supervise the making and installation of the plant and to run it during the earlier years. This and the other early factories restricted themselves to cheese-making. Factory butter-making had to await the invention of the centrifugal separator, which made its first appearance in England at the Kilburn Show. The condensing of milk was a still later development, though there is an account of American condensed-milk factories in the *Journal* for 1872.

The next step was to institute a working dairy as a regular feature of the Show. After experimental efforts at Bristol in 1878, and at Kilburn, the arrangements for Carlisle were placed in the hands of the Aylesbury Dairy Company, who set up a large range of plant, including coolers, weighing machinery, butter-churns and butter-workers, and kept these in almost constant operation during the week. Two of the new separators were also shown at work—" most ingenious machines, but about whose utility a good deal of doubt still exists ; nor is it at all certain that the cream does not undergo some important molecular change from the centrifugal force to which it is subjected." Practical demonstrations and lectures were given throughout the week, and it was clear that the educational results fully justified all the labour and expense involved.

In 1888, at Nottingham, the scope of the dairy section was still further widened by instituting a butter-making competition. Fortunately, as it happened, the entry was small, for an unforeseen vagary of the weather seriously interfered with the progress of the event. Elaborate measures had been taken, in the siting and installation of the dairy, against the risk that the butter would suffer from the heat-wave that Royal Show week so often produced ; but in place of this there came bleak days and a chill north wind, and the dairymaids shivered in their white linen while " the cream took an inordinate time to become butter and fairly puzzled all the professors." Despite its small initial success, the competition aroused the keenest interest, and was obviously a thing to be continued. In fact, the number of entries increased rapidly in subsequent years, and at times became embarrassingly large.

An earlier effort had been made to organise a cheese-making competition, and one trial was actually made in connexion with the Preston meeting of 1885. Here the work was done in a private dairy some time before the Show, and the final awards were made only when the cheese had been fully ripened. There were, however, only two competitors and, since it was evident that it would be a heavy undertaking to run a satisfactory contest on a large scale, the scheme was not continued. With the development of courses of instruction leading up to the National Diploma and other Dairy examinations, the need for cheese-making competition disappeared.

With the expansion of the liquid-milk market and the decline of farmhouse manufactures the Society turned its attention to the possibility of improving the quality of milk and the hygienic standards in its production and handling. In modern times ' Clean Milk Competitions,' organised by County Agricultural Departments, have done a great deal to this end, and the pioneer efforts of the Society are therefore of some historical interest.

The first of these was made in connexion with the Manchester meeting of 1916 when, as might have been expected, most producers were working short-handed and under other war-time difficulties. Nevertheless there was an entry of fifty-nine herds. Samples of the milk, as delivered from each of these, were taken at intervals over a period of a month, were analysed for fat and other solids and were tested, by the crude methods that were then the best available, for ' dirt ' and ' slime.' Points were allotted under each head, with a weighting which gave something like equal emphasis to composition and cleanliness.

By 1923, when the next Milk Competition was held at Newcastle, much better tests of cleanliness had been devised, and the milks were examined by the plate-count method for total numbers of organisms as well as separately for *Bacillus coli.* The standard adopted gave full marks where the total count was less than 30,000 per cubic centimetre and where *coli* was absent from 1 c.c. It is interesting to note that, even in those early days of modern hygiene in the dairy, eleven of the twenty-nine herds scored full marks for cleanliness, so that, in the event, the prizes were awarded mainly on chemical composition.

Another competition, on the same lines as the last, was carried through at Leicester in 1924. There were 22 entries, of which 13 reached the standard for certificates of merit, most of the others failing to obtain the minimum marks under the head of *coli* contamination. The idea of the clean-milk competition was now being widely applied, and it was unnecessary to carry the Society's scheme further.

CIDER AND PERRY

The reader who is interested in the history of cider-making will find an early paper on the subject in the *Journal* for 1843. At that time " no other liquor for ordinary use was thought of in Hereford, Gloucester, Worcester, Somerset and Devon, and it would there have been thought very expensive and troublesome to be under the necessity of supplying its place by brewing malt liquor." But the industry had suffered a decline during the preceding half-century ; the bulk of the produce was consumed on the farm and in the villages and, since neither the farmer nor the labourer was much of a connoisseur, little care was taken to produce a beverage of good quality. In the towns there was a growing preference for beer, which could be had of reliable quality.

A later essay, of 1864, shows that some progress had been made in the science and technique of manufacture, but that the old haphazard methods continued to prevail on most farms. Classes for both cider and perry were provided at the Kilburn show, but the judges still spoke of a declining demand, and evidently considered that this was to be explained, in large measure, by the poor quality of most of the output, as represented by the exhibits.

In 1888, a Mr. Chapman wrote a scholarly article on *Recent Improve-*

ments in Cider and Perry Making, pointing out that, although cider had fallen into disrepute, the little of good quality that was available commanded a profitable market, and that West-Country farmers, in the hard times that prevailed, could no longer afford to neglect such a useful source of income. The author gave an account of the better varieties of fruit, discussed the process of manufacture in scientific terms and pointed out the supreme importance of cleanliness at every stage. Finally, he suggested the establishment of factories as the only means to the production of the bulk supplies of standard qualities which alone would hope to find a profitable market.

At the Windsor Jubilee Show in the following year classes were provided both for grower-makers and for factory owners, and ninety-two samples were sent by thirty-five individuals. Some of these samples were good, but the indifferent quality of many others could not be fully excused by the fact that 1888 had been a poor season. The Council thought that an annual competitive exhibit might do something to raise the standard of quality, and since that time cider has always had its place in the produce section of the Show.

In 1890, when the Society visited Plymouth, the further step was taken of holding trials of cider-making plant, the scene of the event being Glastonbury Abbey. The opportunity was used to demonstrate the whole process, and especially those hygienic measures that were still so often neglected.

It cannot be claimed that the Society played a leading rôle in the revival of the industry which ultimately came about, whereby the popularity of cider was not only re-established in its original district but spread throughout the whole country. That revival was brought about by the combined efforts of the manufacturers themselves, of the Long Ashton Research Station, of the Markets Division of the Ministry of Agriculture, and of the Bath and West and the 'Royal' Societies.

It must suffice to make a bare mention of two of the many new features that have been added to the Shows of more recent times. On several occasions up till 1910 the Society encouraged local horticultural societies to hold their exhibitions in its showyard, but from that year the Flower Show has been organised by the Society itself and has been held annually. In recent years, thanks largely to the work of Sir Arthur Hazlerigg, it has grown until it is now one of the leading annual events of its kind. It has done a good deal to stimulate the interest of the farmer, or at least of the farmer's wife, in the too often neglected farm garden.

An Education and Research Exhibit was first staged in 1903 and, although the early efforts were not calculated to develop any great enthusiasm, the scientist has gradually learnt the art of showmanship, and has found that it is worth while properly to display his wares. A standing committee of the Society, with representatives of the Ministry of Agriculture, now consults with the staff of the Advisory Province in which the Show is to be held, and the thought and labour which the latter have devoted to the exhibits of recent years have been amply repaid by the great interest of the many visitors.

THE SOCIETY'S WAR WORK

One of the duties which the Society has taken upon itself in time of war has been the relief of its fellow-farmers in the devastated areas. Its first effort was made in connexion with the Franco-Prussian War of 1870–71, at the conclusion of which great numbers of French peasants found themselves without seed corn, work animals and other of the bare necessities to bring their land again under crop. The Society opened a subscription list, collected a sum of £52,000 and, through its sister Association in France, was able to relieve a great number of the most needy cases.

In 1914 it became obvious that another and a larger problem of the same kind was certain to arise in several of the allied countries, and early in 1915 a committee was formed to take the necessary steps for relief. Its task was partly to co-ordinate the efforts that were already being made by various individuals and institutions, and partly to enlist the active help of the Society's own members. There was, in fact, some feeling that the effort was premature, and that it would have been better to wait until the nature and size of the problem could be gauged. But it was wisely resolved to begin at once on the task of collecting money, and to lay this aside until it could be usefully spent. A large and widely representative executive committee was formed with the Duke of Portland (the President of the Society) as its head, and with Lord Northbrook as Chairman and Mr. Charles Adeane as Treasurer. His Majesty King George V gave his patronage to the work.

The Executive appointed a number of standing committees to collect and administer the funds, to make purchases of live stock, to provide implements, seeds and other farm requirements, and, latterly, to enlist co-operation of farmers in the Dominions. Branch committees were formed in every county in order to collect money, and especially to organise the gift sales which were a most productive source of funds. The Breed Societies and Smithfield Club also co-operated, and the Highland Society organised a corresponding effort in Scotland.

Up till 1920 a total of £198,000 had been collected in cash, and gifts of live stock, seed, etc., brought the total to £253,000.

During the course of the War some gifts of machinery and seeds were distributed in the reoccupied areas of France, but the main task of the supply committee at this time was to discover how the money could be most usefully spent, and to make plans for the speediest possible action on the conclusion of hostilities. Live stock was found to be everywhere the chief and most urgent requirement.

As soon as possible after the armistice the Committee's gifts were made to the Governments of the various countries which had suffered, were delivered free of charge at conveniently situated seaports and were there handed over to the persons who had been appointed to receive them. The distribution, according to value, was :

France .	. £92,209		Serbia .	. £68,525
Belgium.	. £60,770		Roumania.	. £13,481

The list of actual gifts included 2,700 head of breeding cattle, nearly 3,900 sheep, 1,400 pigs, 400 goats and 19,000 head of poultry ; 3,000 sacks of seed corn, 900 sacks of seed potatoes, 42 tons of vegetable seeds, 29 tons of grass seeds and 9,000 fruit trees ; 900 large implements, from ploughs to threshing machines, and 8,000 hand tools.

The work of distribution necessarily continued for long after the conclusion of peace, for, in many districts, shell-craters and trenches had to be filled in and housing had to be erected before the farmers could think of tillage or of stock-keeping ; it was not till 1923 that the end was reached.

The administrative expenses, in issuing appeals, organising sales, etc., amounted to only some £10,000, and this amount was more than balanced by the interest earned on the temporarily invested funds. The scheme was run upon admirably sound lines, and the gratitude of the great numbers of those who benefited was an ample reward to those who gave so freely of their money, goods and services.

As is well known, the outbreak of the war in 1914 found the country without any plan for increasing the home production of food or for ensuring the equitable distribution of the available supplies, and it was nearly two years before it was realised that an unprecedented effort would be required to avert starvation. In the critical time of the enemy sub-marine campaign a variety of measures had to be hastily improvised—measures to increase the areas of cereals and potatoes, to ration animal feeding stuffs, to control markets, to fix farmers' prices and to avoid waste in every form. To these ends all the agricultural knowledge of the country had to be mobilised, and here again the Society played its part. Its War Emergency Committee on the one hand kept in constant touch with the Government Departments concerned, providing technical information and advice, while, on the other hand, it helped to stimulate the efforts of its members and of farmers in general.

THE SOCIETY'S GOLD MEDAL

As was said in Chapter IX, the Society decided in 1933 to institute a Gold Medal, which it determined to award, at intervals of a year or longer, for distinguished service to agriculture, either in practice or in science, the award to carry with it the Honorary Membership of the Society.

The first choice for the new honour was Sir Thomas Hudson Middleton, who, after a distinguished career in education and research, became successively Assistant Secretary to the Board of Agriculture, Deputy-Director of Food Production (in 1917) and, in 1919, a member and later Vice-Chairman of the Development Commission. In these various capacities his breadth of knowledge and his wise judgment have been of inestimable value to British Agriculture and to the State.

Next in order came the late Lord Ernle (Mr. Rowland Prothero), who was President of the Board of Agriculture during the years 1916–19, when this country's farmers were called upon for the greatest effort in history. But perhaps the selection Committee had in mind, as much as his services as an administrator, Lord Ernle's distinction as a scholar, and particularly his authorship of the classical work on the history of English Farming.

The third recipient was the late Sir Arnold Theiler who, by his brilliant researches in animal pathology " made the Dominion of South Africa habitable by domesticated animals," and whose methods of investigation have proved of immense value to veterinary science through-out the world.

In 1936 the honour was conferred upon Sir William Cecil Dampier, who, after earning a position of distinction in mathematical science and in economics, became the first Secretary of the Agricultural Research Council. The success of this body in the important task of directing and co-ordinating the country's scheme of agricultural research is a tribute to his ability and devotion.

In 1937 the Society selected one of the most active and distinguished members of its own Council, who had already been its President in the previous year. Sir Merrik Burrell is one who, like Spencer and Pusey a century earlier, has given himself to the cause of agricultural progress. Among his many services the greatest has been to carry through the reorganisation of veterinary education, research and administration, to give the veterinary profession the status which it deserves and to prepare the way for a planned attack upon the problem of animal disease.

The award in the Society's centenary year went (and by a happy coincidence along with the honour of Knighthood) to Professor R. G. Stapledon, whose brilliant work on the breeding of herbage plants and on the improvement of pastures would alone have justified both honours. But apart from his solid services as a research worker, Sir George Stapledon's enthusiasm for better farming has been an inspiration to all his fellow-workers in the cause.

CHAPTER XIV

THE CENTENARY SHOW

THE approach of its hundredth anniversary found the Society in a
flourishing condition, and there was every argument for planning
a celebration upon a reasonably generous scale. It was obvious,
too, that the Show should be the occasion, although, as has been mentioned elsewhere, the end of the century was marked by certain other
changes and improvements, especially in the *Journal*.

The only shadow that cast itself over the preparations was that of
the international situation. It was a possibility, and one that seemed to
loom nearer as time went on, that an outbreak of war would bring the
Society's plans to nothing, and involve it in a loss of anything up to
£30,000. However, this risk was one that the Society could afford to
bear ; even if the worst had happened seven-eighths of its reserve fund
would have remained, and none of its activities need have been seriously
hampered. In the event, of course, the country was at war before the
dismantling of the Showyard had been completed.

His Majesty the King was graciously pleased to become the Society's
President for the year, as Queen Victoria had been for the year of the
Jubilee. His Majesty took the closest interest in the year's proceedings
and, in particular, did a great deal to ensure the success of the Show.
He appointed the Earl of Athlone to be his Deputy, and the Earl was
most assiduous in his attendance at Council meetings, and in his efforts
to make the Show a really notable one.

The first problem in connexion with the Show was the choice of a
site, and consideration of this began six years in advance. One possibility, attractive at least on sentimental grounds, was to return to Oxford,
where the founders had gathered for the first Country Meeting. But the
area of land required for 1939 was nearly twenty times as large as what
had sufficed a century earlier, and no conveniently situated site of this
size could be found near the city. Another thought was to pitch the
Show near some large industrial town of the Midlands or the North where,
according to all past experience, the maximum attendance might have

LORD DARESBURY

HONORARY DIRECTOR OF SHOWS,
1906–30

SIR ROLAND BURKE

HONORARY DIRECTOR OF SHOWS,
1931–39

The Showyard, Windsor, 1939

been expected ; but it was generally agreed that a financial success was not the prime consideration. The third proposal, and that which received the most support, was that the Society should beg leave to return to the scene of its Jubilee meeting in Windsor Great Park. Accordingly, in 1933, it was arranged that the Duke of Devonshire, the President of the year, should discuss the matter with King George V on the occasion of His Majesty's visit to the Derby Show. His Late Majesty received the suggestion with the greatest cordiality and readily gave his consent. The arrangement was confirmed, in turn, by King Edward VIII and by King George VI.

Windsor Great Park proved, in every way, an admirable setting. The site, indeed, required a good deal of drainage as an insurance against the risk of a wet show week, and it was also true that the approaches to the ground were barely adequate to carry the streams of cars on the busiest days. But against these minor drawbacks there were the practical advantages of ample elbow-room and of a level expanse of strong old sward. Moreover, the fine old trees scattered over the area relieved the formality of the lay-out, without being numerous enough to make any practical difficulty for the surveyor. Indeed, despite the unprecedented number of exhibits and the extra width allowed for the avenues, the arrangement of the yard was conveniently compact, the distances from point to point being less than they have sometimes been at an ordinary Show. The area of the Windsor Showyard, excluding car-parks, was 126 acres.

Fortunately, the two individuals[1] upon whom rested the main responsibility for the arrangements were already fully experienced in the organising of the Show. Mr. T. B. Turner, who had been Secretary of the Peterborough Agricultural Society, succeeded Mr. McRow as Secretary of the 'Royal' in 1921, and thus had nine years under Lord Daresbury's Honorary Directorship. Mr. U. Roland Burke began work for the

---

[1] The list of Show officials was as follows : Honorary Director, U. Roland Burke ; Stewards of Horses, Cyril E. Greenall, O.B.E., and Thomas Forshaw ; Steward of Ring, Major Gordon B. Foster ; Stewards of Cattle and Goats, Walter R. Burrell and Hugh Tinsley ; of Sheep and Pigs, John F. Harris and F. G. B. Stephens ; of Veterinary Examinations, Cyril E. Greenall, O.B.E. ; of Implements, John Bell ; of Dairy, Poultry and Produce, James Mackintosh, O.B.E. ; of Forestry, Lord Hastings ; of the Flower Show, Sir Arthur Hazlerigg, Bt. ; for Reception of Overseas Visitors, Sir Merrik R. Burrell, Bt., and Lt.-Col. Sir Archibald Weigall, Bt. ; of Catering, John Bourne ; and of Finance, Charles Adeane, C.B., Sir L. Foster Stedman and Col. J. C. H. Wheatley ; Chief Veterinary Officer, Major Brennan De Vine, M.C., F.R.C.V.S. ; Consulting Engineer, S. J. Wright, M.A. ; Surveyor, Charles H. R. Naylor, L.R.I.B.A. ; Secretary, T. B. Turner.

Society as an assistant steward as early as 1893, under Mr. Cecil Parker, and, after graduating through various departmental stewardships, succeeded Lord Daresbury as Honorary Director in 1930. In the succeeding years the Show, all things considered, reached the highest standard in its history. Most members will recall especially the succession of Derby in 1933, Ipswich in 1934 and Newcastle in 1935, when the Society's own efforts were supported by those of most enthusiastic and efficient Local Committees and when the sun shone, with unusual consistency, upon their united labours. The following three Shows were marred, more or less, by rain, but were hardly less excellent in themselves. The finest collection of live stock was at Ipswich, while the greatest exhibit of machinery was staged at Wolverhampton in 1937, the year when Mr. Burke occupied the double office of President and Honorary Director. Indeed, the only unsuccess of the decade was the Southampton Meeting of 1932, when small entries were followed by the poorest attendance since Park Royal. With this sole exception the recent Shows had been so large, so well organised and so successful that it was hard to see how and where improvements were to be made. This was the main difficulty that faced a Special Committee appointed, in 1937, with instructions to make the Centenary Show something quite unique. That they succeeded was generally admitted by the many show-goers whose memories stretched back to the Jubilee and beyond.

LIVE STOCK

The first steps necessary to ensure a large and fully representative exhibit of our British breeds was to provide an adequate Prize Fund, and to make a place in the classification for every breed that seemed to be capable of producing adequate numbers. The prizes for live stock (excluding produce, etc.) reached an aggregate value of £17,056, to which the various Breed Societies contributed a total of £3,752. The Racecourse Betting Control Board also gave £1,000 towards the prizes for horses. The balance (£13,304) came from the Society's own resources. The total Fund was about £6,500 more than that provided for the Jubilee Show. Special Centenary Gold Medals were struck, to be awarded to the winners of first prizes in the inspection classes for horses, cattle, sheep and pigs ; and, of the four Champion Cups presented by His Majesty, three were allotted to live-stock exhibits, the fourth going to the Flower Show. The King's Cups for live stock were awarded (1) for the best riding hunter, (2) for the dairy cow which, having won a first, second or third prize in its inspection class, gained the highest number of points in the milk-yield competition, and (3) for the best dairy or beef calf shown in the Young Farmers' Clubs' classes.

The table on the following page sets out the classification, and gives, for each breed, the number of entries. The corresponding figures for the Jubilee Show are also given. The total entries of horses, cattle, goats, sheep and pigs were 4,548, a figure which compares with 3,997 at Windsor in 1889, with 2,874 at Kilburn in 1879, and with 3,534 at Ipswich in 1934—which last was the biggest Show, as regards live stock, of recent times. The Centenary exhibit was thus the largest that had ever been brought together in this country. It failed in completeness only in respect of the mountain breeds of sheep, the absentees being the Blackfaced Mountain, the Cheviot and the Exmoor Horn.

As will be seen, a few breeds that were of considerable importance fifty years ago—the Hackney and Hackney Pony, the Cotswold Sheep and the Small White Pig—had dropped out of the list. All of these, in their several ways, represented considerable achievements of the breeder's art, but all have been condemned to virtual extinction by changes in our ways of life or in our taste in the matter of meat. Of the newcomers the Percheron, the Arab and the Friesian are naturalised aliens ; others, such as the Cumberland, Essex and Wessex Pigs, are not truly new, but are old local types which have been ' fixed,' improved and brought under pedigree registration only in comparatively recent times. Still others— the Dairy Shorthorn, the Dorset Down and the Black Welsh Mountain —are branches of old-established breeds and not new creations. The list of new breeds in the strict sense of the words—of breeds that have been created by processes of crossing and re-selection—is thus very short ; probably it should include only Blue Albion Cattle and Devon Closewool Sheep. This is not to say that the last two generations of breeders have been less enterprising than their predecessors. There can be no point in making a new breed unless there is a definite place and a clear need for it ; most of the places had been filled fifty years ago, and most of the breeds of that time have proved plastic enough to meet the gradually changing needs, and to be adapted to changing conditions of farming.

Some of the broad changes in our live-stock industry are reflected in a comparison of the figures for the two Windsor Shows, though we must remember that the real importance of a breed is not necessarily to be measured by the number of Royal Show entries which it produces. This number is influenced, in fact, by many things. One is the extent

THE CENTENARY SHOW

COMPARATIVE STATEMENT OF ENTRIES, Etc.

At the Shows held at Windsor in 1889 and 1939

HORSES, CATTLE, AND GOATS.	1889.		1939.		SHEEP, PIGS, AND POULTRY.	1889.		1939.	
	Classes	Entries	Classes	Entries		Classes	Entries	Classes	Entries
HORSES :—					**SHEEP :—**				
Prizes		£3,038		£4,781	*Prizes*		£2,507		£2,372
Shire	9	167	12	90	Oxford Down	4	82	5	47
Clydesdale	9	93	6	50	Shropshire	4	212	6	25
Suffolk	9	105	13	168	Southdown	4	123	7	97
Percheron	—	—	12	109	Hampshire Down	4	78	5	58
Hunter	11	279	11	141	Suffolk	4	35	9	81
Polo Pony	—	—	5	47	Dorset Down	—	—	4	44
Cleveland Bay, Coach Horse	7	57	2	18	Dorset Horn	4	31	3	25
Hackney	11	148	—	—	Wiltshire Horn	—	—	3	14
Pony	8	67	—	—	Ryeland	3	15	4	29
Arab	—	—	4	33	Kerry Hill (Wales)	—	—	5	32
Welsh Pony	—	—	4	17	Clun Forest	—	—	5	33
Shetland Pony	—	—	5	35	Lincoln	4	58	5	31
Riding Classes—					Leicester	4	41	4	24
Hunters	—	—	8	161	Border Leicester	4	31	6	40
Cobs	—	—	1	15	Wensleydale	3	19	5	19
Hacks	—	—	3	54	Kent or Romney Marsh	3	36	6	34
Riding Ponies	—	—	5	56	Cotswold	4	60	—	—
Children's Ponies	—	—	3	55	Devon Long Wool	3	31	2	6
Driving Classes	4	41	11	37	Devon Close Wool	—	—	2	6
Jumping	—	—	5	315	South Devon	—	—	2	10
Mounted Police	—	—	2	27	Dartmoor	3	12	2	8
Draught Horses	4	15	—	—	Hardwick	3	33	—	—
Asses	2	17	—	—	Exmoor Horn	3	20	2	—†
					Roscommon	3	14	—	—
Total for HORSES	74	989	112	1,428*	Limestone	3	10	—	—
					Lonk	3	17	2	8
CATTLE :—					Blackfaced Mountain	3	26	3	—†
Prizes		£3,952		£7,379	Cheviot	3	27	2	—†
Shorthorn	8	222	11	83	Welsh Mountain	3	34	4	31
Hereford	6	121	10	63	Black Welsh Mountain	—	—	3	26
Devon	6	84	6	36	Any other British Breed	2	24	—	—
Sussex	6	97	4	29					
Welsh	6	49	4	19	Total for SHEEP	81	1,069	106	728
Park	—	—	3	15					
Longhorn	2	11	4	15					
Aberdeen-Angus	6	87	9	151	**PIGS :—**				
Belted Galloway	—	—	5	48	*Prizes*		£740		£1,917
Galloway	6	46	4	29	Large White	4	27	9	269
Highland	2	18	3	20	Middle White	4	31	9	89
Dairy Shorthorn	—	—	11	253	Small White	4	23	—	—
Lincolnshire Red Shorthorn	—	—	7	55	Tamworth	4	45	7	48
South Devon	—	—	5	31	Berkshire	4	96	9	55
Red Poll	6	71	10	160	Wessex Saddleback	—	—	9	117
Blue Albion	—	—	5	17	Large Black	4	43	8	97
British Friesian	—	—	13	207	Essex	—	—	8	130
Ayrshire	6	50	9	131	Gloucestershire Old Spots	—	—	6	35
Guernsey	5	141	9	175	Cumberland	—	—	6	22
Jersey	7	434	8	222	Long White Lop-Eared	—	—	5	22
Kerry	3	77	4	15	Welsh	—	—	6	27
Dexter	3	59	6	45					
Any other Breed	2	8	—	—	Total for PIGS	24	265	82	911
Milk Yield	6	62	11	142					
Butter Test	—	—	2	101	**POULTRY :—**				
					Prizes		£334		£346
Total for CATTLE	86	1,637	163	2,062*	Entries	89	862	100	522
GOATS :—									
Prizes		£45		£140					
Inspection Classes	6	37	12	114					
Milk Yield	—	—	2	100					
Total for GOATS	6	37	14	214*					

* Animals exhibited in more than one class are here counted as separate entries.
† Classes cancelled under regulations of Prize Sheet.

of the export demand for the time being, or perhaps the prospects, in the opinion of the breeders, of stimulating this demand ; for instance, the apparent decline of the Shropshire since 1889 is to be largely explained by the fact that there was a very large American market for the breed about the time of the Jubilee. Again, certain breeds, such as the Jersey and the Southdown, have always proved attractive to the wealthier classes, who can afford to show upon a large scale. Thirdly, even with the great improvement in transport facilities for stock, the distance between the particular Show centre and the home area of a breed is still a consideration ; North-Country breeds tend to be poorly represented at South-Country Shows, and *vice versa.*

It is true, however, as the figures would suggest, that our pig industry has greatly expanded during the past fifty years and that the rôle of the sheep, in the main arable districts, has become less important ; that cattle breeding was quite as important in 1939 as in 1889, but that the emphasis had shifted from beef to milk ; and that the trend of public demand in the meat market had favoured such types as the Large White, the Suffolk Sheep and the Aberdeen Angus, where the production of meat of high quality, with the absence of excess fat, has been the main object of the breeders.

As regards the standard of quality of the stock at the Centenary Meeting, it was natural that breeders and Breed Societies should make very special efforts on the occasion, and the result was that, upon the whole, the level of merit was the highest in the Society's history. Since the Stock Catalogue was a book of 500 pages, it is clearly impossible to describe the classes each by each, and all that will be here attempted is to single out some of the most remarkable of these.

The breeding classes for Draught Horses were generally good, with the Suffolks, as has commonly happened in recent years, providing the highest entry. The outstanding feature of the section was, however, the display of four-horse teams in harness. Here, again, the Suffolk was most largely represented, though many horsemen thought that the first-prize Shire team, of four really magnificent greys, was the most remarkable of all. On the other hand, it was one of the Suffolk teams, parading in old-time harness with its full complement of brasses and bells, which was the favourite with the ring-side crowd. Several exhibitors had brought extra pairs and were thus able to parade teams of six.

The hunter classes were both large and good, ponies were well represented and there were some very beautiful Arabs. The old and handsome Cleveland Bays provided another very interesting group.

As at several recent Shows, a special exhibit of colliery horses and pit ponies was arranged by the Mining Association of Great Britain. This included representative animals from all the main coal-fields, ranging in size from a nine-hands pony from Northumberland to a group of heavy Shires from South Wales. Many of the animals had long periods of service underground, and all were shown in perfect health and condition.

Turning to the cattle, it must be admitted that several of the beef breeds turned out in disappointingly small numbers. The notable exception was the Aberdeen Angus, which produced 151 entries in its nine classes, with a very high standard of quality and a remarkable uniformity of type throughout. The breed had never been seen to such advantage at a Royal Show.

If some of the beef classes disappointed a little, those of the dairy and dual-purpose breeds more than fulfilled the highest expectations. Numbers of Dairy Shorthorns, Friesians, Jerseys and Guernseys were almost embarrassingly large, and it was often hard indeed to distinguish degrees of merit, so that judges were at work until very late in the day. The Red Poll exhibit was also remarkable, though it may hardly have equalled the famous one at Ipswich five years earlier. The Ayrshires were by far the finest collection ever seen at the ' Royal.' The most remarkable class in the whole section, however, was that specially arranged for the occasion by the British Friesian Cattle Society and confined to cows of the breed which had given, in a lactation period as defined by the Society's rules, a minimum of 2,000 gallons of milk. Nine cows paraded, all in the bloom of perfect health. Indeed, the winner showed, by winning also the female championship of the breed, that perfection of form is by no means incompatible with the highest productivity.

Ninety-six cows competed in the milking trials and yields were high, the breed averages ranging from 76 lb. for Lincolnshire Red Shorthorns to 36 lb. for the tiny Dexters. The highest score was that of a Friesian which gave 97·75 lb. of milk, containing 3·55 per cent. of butter fat, in the twenty-four hours. This cow, however, was not qualified to compete for the King's Cup ; the latter was won by a representative of the same breed with a yield of 88·4 lb. and a butter-fat percentage of 4·04. Still another Friesian—the winner in the two-thousand-gallon class—stood Reserve.

Another interesting event was the Cattle Show of the Young Farmers' Clubs, which took place on the Friday of the Show week. The first show of this kind took place at Newcastle-on-Tyne in 1935, but in that year, and in the three succeeding ones, the classes were restricted to the locality of the Meeting. At Windsor there were both local and open classes, and the event attracted a total of 200 entries. The quality and the condition of the stock made a high testimonial to the skill of its owners.

As has been said, the sheep section was the only one that failed to attain completeness, and in some of the breeds that appeared numbers were smaller than could have been wished. Nevertheless, twenty-three breeds, with 728 entries (including, of course, a large number of pens of three animals), was the best display that most members had ever seen. The Southdown and the Suffolk, with 97 and 81 pens respectively, led in the matter of numbers. His Majesty (who was also a successful exhibitor of Devon and Red Poll cattle) sent Southdowns from his Sandringham flock, and had first prizes for both shearling rams and ram lambs. An interesting non-competitive exhibit was a collection of Shetland sheep, which provide the raw material for the well-known shawls and other hand-knitted goods of the Islands.

The pig exhibit was affected to some extent by the rather widespread outbreak of swine fever in western districts, but it could be said of at least five of the eleven breeds that numbers were as large as could reasonably have been wished, and that, throughout the whole section, the standard of quality was higher than it had ever been. Indeed, the five leading breeds—the Large White, Middle White, Essex, Wessex Saddleback and Large Black—formed a collection just as notable, of its kind, as that of dairy cattle. The practice of parading the prize-winning animals, class by class, on the later days of the week, which was begun in 1937, was specially appreciated at Windsor, where the judging day was so crowded with interest. A new class, for sow with litter, was provided in the Large White, Middle White, Berkshire and Wessex Sections. The litter, of less than eight weeks of age, was required to be at least eight in number ; judging was by points, 40 being allotted as a maximum for the sow and 60 for her progeny. There was good competition in the case of the first and last of these breeds.

As had been the practice for a good many years, the Young Farmers' International Dairy Cattle Judging Competition was decided in the Showyard on the Wednesday of the Show week. Unfortunately, neither the United States nor any of the Dominions was able to send a team, but England, Scotland, Wales and Northern Ireland all took part. A member of the Scottish team, which won the Gold Challenge Cup, created a record by placing each of the six classes of cattle in precisely the order of merit that had been agreed by the bench of judges.

The Dairy Department, like most others, was on a larger scale than usual. One of its tasks was to receive daily about 1,000 gallons of milk and to convert this into butter, scalded cream and cream cheese. Some five hundred milk samples, goats' milk included, were tested in connexion with the milking trials. Seventy-three cows—a record number—competed in the Butter Test Trials, and the milk of each had, of course, to be churned separately under standardised conditions. The Butter-Making Competitions attracted an entry of 155 and the decisions, in some cases, were very difficult to make. The Inter-County Team Championship, which, as usual, was the most keenly contested event, went to Devonshire. On the last day of the week the Individual Championship attracted even more interest than usual, since the prize was a piece of plate presented by Her Majesty the Queen. The prizes were presented by Lady Burke.

IMPLEMENTS AND MACHINERY

The Implement Yard, with its 430 stands and its mile and three-quarters of shedding, was considerably larger than those at recent Shows, but actually smaller than that of 1889. The comparison is here, however, scarcely valid, on account of the changes that have since occurred —particularly the tendency for the manufacture of implements, as well as the trade in fertilisers, feeding stuffs and seeds, to be carried on by fewer and larger firms. In any case there was a very complete range of all kinds of farm machinery.

Few shows can have produced a longer list of novelties. Entries for the Society's Silver Medal numbered ten, and there were four awards. These went respectively to a new type of swath-turner specially designed for use on ridged or otherwise uneven land ; a sugar-beet harvester which combined the operations of lifting, topping and cleaning the roots ; a sterilising chest for dairy utensils ; and a new type of rubber tractor-tyre designed to prevent wheel-slip on wet ground.

Apart from the medal entries there were several new models of tractors of established makes, and a completely new British tractor with many interesting features. A " baby " combine harvester, a selection of new collecting and loading machinery for hay and green crops, a mower with an endless band in place of the standard reciprocating knife, and a labour-saving machine for distributing grass seed and fertilisers in one operation, were some among the many other inventions of the year.

An exhibit of great historical interest was staged by Messrs. Ransomes, Sims and Jefferies, the firm which, it will be remembered, was awarded

XXVIA.—SHIRE FILLY, " CHENIES MAVIS "
Winner of Champion Gold Medal for best Shire Mare or Filly

XXVIB.—CLYDESDALE MARE, " LUCINDA "
Winner of Champion Silver Medal for best Clydesdale Mare or Filly

XXVIIA.—Suffolk Mare, " Laurel Golden Girl "
Winner of Champion Prize for best Suffolk Mare or Filly

XXVIIB.—Percheron Mare, " Holme "
Winner of Challenge Cup for best Percheron Mare or Filly

XXVIIIA.—SHORTHORN BULL, "MUIRSIDE RAMSDEN KING"
Winner of Champion Prize for best Shorthorn Bull and "Brothers Colling" Challenge Cup for best Shorthorn

XXVIIIB.—HEREFORD BULL, "ASTWOOD DANDY"
Winner of Champion Prize for best Hereford Senior Bull and Challenge Trophy for best Hereford Bull

XXIXᴀ.—Sᴜssᴇx Cow, " Loᴄᴋ Kɴᴇʟʟᴇ 2ɴᴅ "
Winner of Champion Silver Medal for best Sussex Cow or Heifer, and Challenge Cup for best Sussex

XXIXв.—Aʙᴇʀᴅᴇᴇɴ-Aɴɢᴜs Cow, " Eᴜʟɪᴍᴀ 6ᴛʜ ᴏғ Kɪʟʜᴀᴍ "
Winner of Champion Gold Medal for best Aberdeen-Angus

XXXa.—Dairy Shorthorn Cow, " Aldenham Florentia 8th "
Winner of Champion Prize for best Dairy Shorthorn Cow or Heifer

XXXb.—Red Poll Cow, " Wissett Nonsuch "
Winner of Champion Prize for best Red Poll Cow or Heifer

XXXIA.—British Friesian Cow, "Terling Breeze 34th"
Winner of Champion Prize for best British Friesian Cow or Heifer

XXXIB.—Ayrshire Cow, "Garston Orange Blossom"
Winner of "Cowhill" and "Oldner" Challenge Cups

XXXIIA.—Jersey Heifer, " Everdon Fancy's Dream "
Winner of Champion Prize and " Trent " Challenge Trophy

XXXIIB.—British Friesian Cow, " Lavenham Lilac 9th "
Winner of His Majesty the King's Champion Cup

XXXIIIb.—Shropshire Shearling Ewes
Winners of " Hardwicke " Challenge Cup

XXXIIId.—Hampshire Down Ewe Lambs
Winners of Champion Prize for best Hampshire Down Sheep

XXXIIIa.—Oxford Down Shearling Ewes
Winners of " Kelmscott " and " Birdlip " Challenge Cups

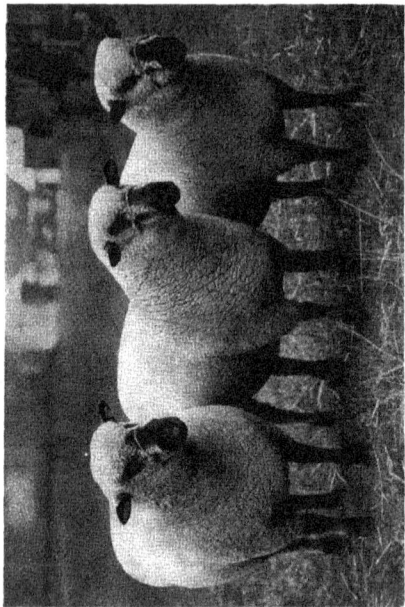

XXXIIIc.—Southdown Shearling Ewes
Winners of Champion Silver Medal and " Northumberland " Challenge Cup

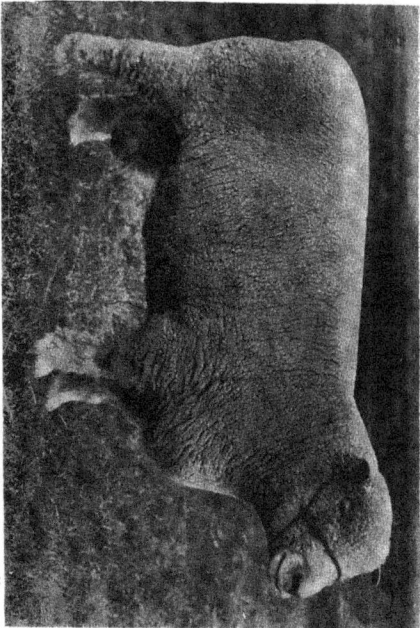

XXXIVa.—RYELAND SHEARLING RAM, "THOMAS'S WEAPON"
Winner of Challenge Cup for best Ryeland Shearling Ram

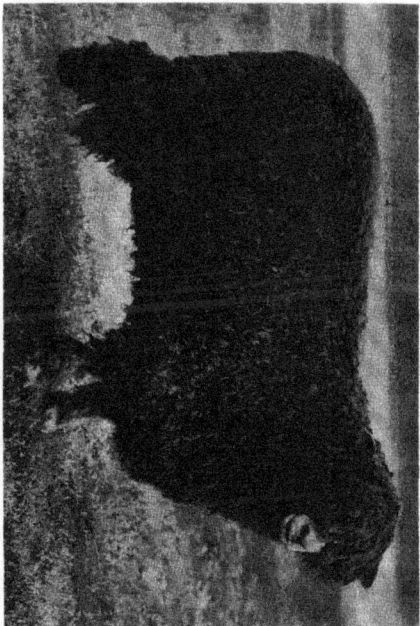

XXXIVb.—LINCOLN SHEARLING RAM
Winner of Special Prize for best Lincoln Shearling Ram

XXXIVc.—LEICESTER RAM

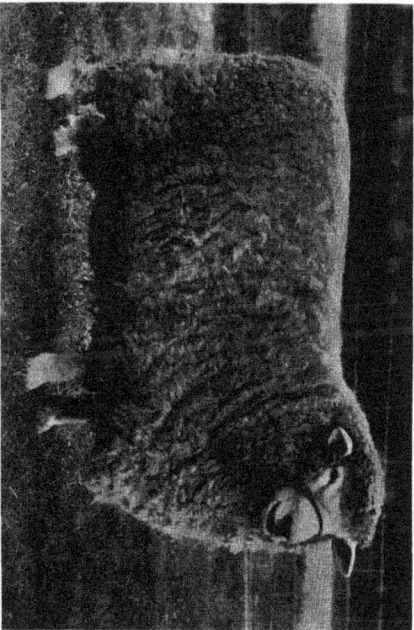

XXXIVd.—KENT OR ROMNEY MARSH SHEARLING RAM

XXXVb.—MIDDLE WHITE SOW, "DUNSDALE PRINCESS 5TH"
Winner of Challenge Cup for best Middle White Pig

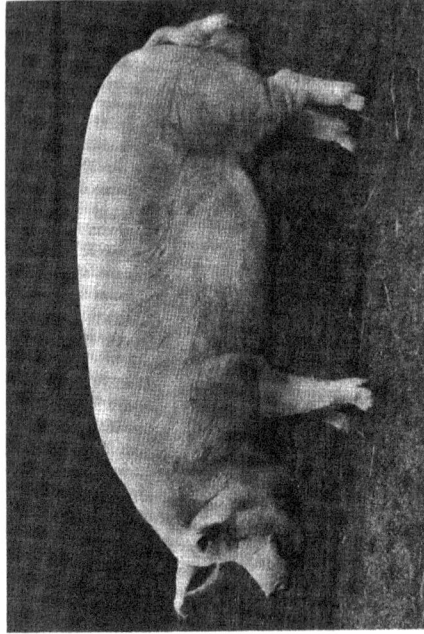

XXXVa.—LARGE WHITE BOAR, "HISTON MOLLINGTON KING"
Winner of Challenge Cup for best Large White Pig

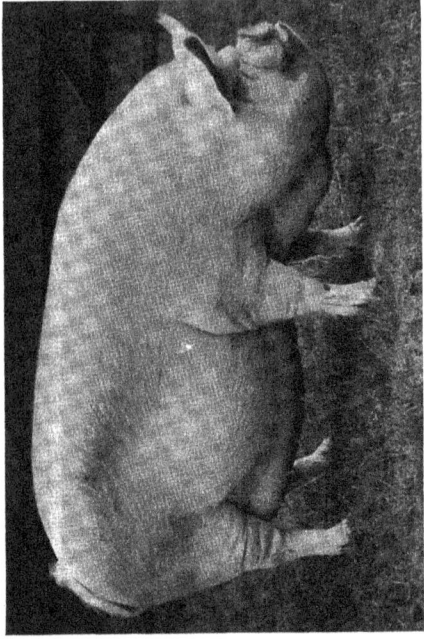

a Gold Medal for its leading exhibit of implements at the first Country Meeting. It remained, at the end of the hundred years, the sole survivor of the implement exhibitors at Oxford, and it had brought a selection of its products to every 'Royal' in the interval. By a happy coincidence Ransomes was celebrating the hundred-and-fiftieth anniversary of its own foundation, and it staged exhibits illustrating both its own history and that of the Society. Among the latter was the Society's only copy, lent for the occasion, of the catalogue of the Oxford Show.

HISTORICAL EXHIBITS

Through the good offices of the Earl of Athlone, who, besides being Deputy President of the Society, was President of the National Horse Association of Great Britain, it was arranged that the two bodies should collaborate in arranging a display of historic horse-drawn vehicles. His Majesty the King graciously lent a number of very interesting carriages and also twenty horses from the Royal stables, with their harness and trappings, so that the vehicles, properly horsed and equipped, could be paraded in the ring. The exhibit was organised by Major H. Faudel Phillips (Honorary Director of the National Horse Association), Colonel Sir Arthur Erskine (Crown Equerry) and the Society's Honorary Director.

In all, forty vehicles were assembled, together with a highly interesting collection of harness, saddlery, horse brasses, etc., much of which was loaned by the Worshipful Company of Loriners. The carriages, apart from those of His Majesty, came partly from public collections and partly from private owners. Twenty-six were paraded, fully horsed, each afternoon.

Among the once common standard types of vehicle were early forms of the phaeton, cabriolet, hansom cab, carrio, pony chaise and 'growler.' Of special historical interest were the original Brougham, an early York mail-coach and Shillibeer's original London omnibus, complete with cockney conductor. Among items with interesting personal associations were state carriages which had belonged to Earl Spencer and Lord Daresbury, a remarkable shooting charabanc of King Louis Philippe of France, Florence Nightingale's Crimean carriage and that in which the Emperor Napoleon III drove to his surrender after the Battle of Sedan. From the Royal collection came King George Fourth's post-chaise ; a state landau, an ivory-mounted phaeton and a garden chair of Queen Victoria, the last appropriately drawn by a Highland pony ; a sledge designed and driven by the Prince Consort ; a town coach of King

Edward VII ; a donkey barouche presented by Queen Adelaide to the Royal children in 1846 ; and a miniature landau presented by the showmen of England to the children of King George V.

More than 36,000 visitors made the round of the exhibit, and the parade constituted a most popular item of each afternoon's programme.

Another of the special exhibits which deservedly attracted a large stream of visitors was the *Old English Farm* organised and arranged by Mr. E. J. Rudsdale of the Colchester and Essex Museum, and containing material of great historical interest.

The buildings consisted of a farm-house flanked by two barns, the whole being built with white-washed weather-boarded walls and with thatched roof. In front was a small old-fashioned garden, and a paved dairy court complete with a leaden pump. The centre portion consisted of a kitchen, dairy and brewhouse all equipped with utensils of the eighteenth or early nineteenth century ; the kitchen had its turnspit, its village-made chairs and tables, its wooden platters and its pewter pots ; the dairy had its plunger-type churn and its hand butter-working utensils.

In the machinery barn was an excellent collection of early hand tools, an interesting series of ploughs of the period about 1800, and a number of horse implements including Bell's and McCormick's reapers.

There were daily demonstrations of butter-making and of the threshing of corn with the flail, followed by the winnowing of the grain by barn fans.

The live stock consisted of a pair of heavy horses and a cob (their harness being fully furnished with brasses), and a pair of Longhorn cattle.

The main contributors to the collection of implements and furniture were the Science Museum, London, and the museums of Colchester, Cambridge and Hereford.

EDUCATIONAL EXHIBITS

As had been customary for several years, separate pavilions were set aside at Windsor for Agricultural Education and for Forestry ; both of these, in keeping with the occasion, were planned upon a larger scale than usual. An exceptional feature of the Centenary Show was a Schools Section, arranged by a committee composed of the Education Officers of the counties in the vicinity of Windsor, and of Windsor itself.

The Agricultural Education Exhibit was the joint work of the Faculty of Agriculture and Horticulture of the University of Reading (including the National Institute for Research in Dairying and the British Dairy Institute) and of the County Agricultural Staffs of Buckingham and Northamptonshire. Buckingham Producers (an association of market growers), in co-operation with the County Horticultural Staff, added a very effective display of fruit and vegetables.

The main object of the exhibit was to illustrate the changes in British agriculture, and the influence of scientific discovery upon its progress, during the century of the Society's existence. The historical note was struck in the forecourt, which, although primarily intended to form an attractive setting for the pavilion, was also used to draw a contrast between the styles of flower-gardening of 1839 and 1939. The old garden was in the elaborate and formal style of carpet bedding, making use of vast numbers of Sedums and other low-growing plants.

The changes in our agriculture—in land utilisation and in live-stock husbandry—were shown in a number of panels, the comparison being, in each case, between 1867 (the earliest year for which statistics are available) and 1938. These panels brought out very clearly the decline of arable farming and, more especially, the fall in the areas devoted to corn and roots ; but they also showed how these declines had been compensated, more fully than is commonly realised, by the increased production of potatoes and vegetables and by the remarkable growth of the dairy, pig and poultry sections of the industry.

The development of dairying was set out in detail by the same means, and a large central panel called " Landmarks in Dairying " pointed to such great steps of scientific progress as the discovery of bacteria by Pasteur and the researches in nutrition upon which was based the modern system of rationing. Modern work on cowshed hygiene, on feeding, and on the value of milk in human nutrition, was also skilfully set out.

The soils and crops section illustrated the progress of soil surveying and mapping, and the value of modern methods of soil analysis as a basis for scientific soil improvement. This was linked on to a graphic explanation of the advisory work that was being done in the Reading Province in connexion with the Government's Land Fertility Scheme, and this, in turn, to the wider problems of grassland improvement and the conservation of grassland herbage. Another item of much interest was a collection of living specimens of the leading varieties of cereals of a hundred years ago. Many of these, of course, have long gone out of general cultivation, but some few, like Rivet Wheat and the Grey Winter Oat, have withstood the competition of the modern products of the plant breeder.

The Poultry exhibit brought out both the immense expansion of the industry, especially in the past twenty-five or thirty years, and the revolutionary changes from the old natural systems of hatching and rearing to the intensive, mass-production methods of the present day, involving the use of mammoth incubators, battery brooders and laying batteries, and necessitating scientific control at every stage.

The Pig section was the work of the Northamptonshire Farm Institute and was planned to show the value of pig recording and litter testing as aids to improvement. This it did in a very convincing way.

The progress of Veterinary Science, and the benefits accruing to the farmer from veterinary research, were illustrated by setting out the history of three major stock diseases.

There was a striking fruit demonstration, contrasting the results from neglected trees with those obtained by means of scientific management, including the correct choice of root stocks, due attention to spacing and pruning and a complete system of pest control.

The exhibit of choice-quality salads, vegetables and soft fruits, staged by local growers and all packed for market by the most approved methods, was not only an object-lesson to other growers, but must have done a great deal to stimulate the demand of the general public for fresh home-grown produce. The whole of the material had, of course, to be replaced each day during the Show.

The Forestry Exhibition was the largest ever staged, the pavilion and the outdoor exhibits together occupying an area of an acre and a half. The ten competitive classes for gates and fences brought out material that reached a very high level of craftsmanship. The non-competitive exhibits were more numerous and finer than at any previous Show; they included one on the nursery technique for cricket-bat willows, others on the burning of charcoal and the control of forest fires and a collection of axes going back to the Stone Age. Special efforts had been made to demonstrate old-time rural crafts, the list including reed-thatching, wood-turning on the pole lathe, hurdle-making, oak-spelk basket-making and the hand forging of Sussex tools. The Society had the valued assistance of the Royal English Forestry Society in the organisation of the exhibit, and five silver medals were awarded as tokens of gratitude to the various private individuals and public bodies who contributed items of special interest.

The Schools Section was admirably designed to demonstrate what is perhaps the greatest advance in modern school education—the movement away from purely book learning and towards a system which, through the introduction of practical subjects, makes the school more truly a training ground for life. The Rural Science exhibit showed how pupils could be helped, through school gardening and practical work in science, to understand nature and country affairs, and to appreciate the contribution of scientific research towards greater efficiency in food production. The Health exhibit showed what the school could do to lay the foundations of physical fitness, and impart a sound knowledge of nutrition and nutritive values.

Arts and crafts—wood and metal work, spinning, weaving and needlecraft—were, naturally, given a great deal of space, but not to the exclusion of workaday things such as butter-making, poultry-work and cookery. Architecture, music and the drama were also shown to have their place in the modern curriculum.

THE FLOWER SHOW

The efforts of the Flower Show Committee, under Sir Arthur Hazlerigg's chairmanship, resulted in an exhibition which was not only by far the finest that the Royal Agricultural Society had ever produced but which was also, in the opinion of many visitors, as fine as anything of the kind ever achieved in this country ; at any rate, coming several weeks after the time of the Chelsea Show of the Royal Horticultural Society, it brought forth the midsummer flowers, such as Delphiniums, in unexampled perfection.

The tents were arranged to form three sides of a square ; the open-fronted quadrangle so produced was divided between a number of exhibitors, yet was so planned as to form a very pleasing whole. The entrance, a finely designed lych-gate with Norfolk reed thatching, completed the impression of an old-established garden. Inside the tents was a long succession of exhibits in great variety and each magnificent of its kind. The Judges who were given the task of selecting, for the King's Cup, the best among so many had indeed a difficult task. The large number of visitors, and the great pleasure which they obviously found in their visits, left no doubt that the organisers were rewarded as they had hoped to be.

THE WEEK'S PROGRAMME

The judging of the breeding classes of live stock was carried out, as usual, on the first day (Tuesday, July 4th). There had been early rain, but the day was for the most part sunny and pleasant. Arrangements worked smoothly although, as has been said earlier, some of the breeds were so largely represented that the awarding was not completed till late in the afternoon. His Royal Highness the Duke of Gloucester paid an informal visit, and there were many parties of farmers from overseas. These included a large number from the Dominions and also a delegation of fifteen members of the Société des Agriculteurs de France. This party had arrived on the previous Saturday, and had made a tour of Sussex and a visit to Cambridge. They returned to the Show on the Wednesday, when they were presented to Their Majesties. The attendance of over 7,000 was the largest, for the opening day, since 1920.

On Wednesday, when again bright sunshine followed early showers, Their Majesties the King and Queen, accompanied by the Minister of Agriculture, Sir Reginald Dorman Smith, arrived in the Royal Carriage, drawn by the Windsor Greys. They were conducted to the Royal Pavilion and were received by the Deputy President and Princess Alice. Their Majesties visited the Flower Show, spent some time in the Royal Box watching events in the main ring and inspected the Colliery Horses and the Forestry Pavilion.

After luncheon on this day the King conferred the honour of Knighthood upon the Honorary Director, as Queen Victoria had done on the occasion of the Jubilee Show. The Deputy President conveyed to Sir Roland Burke the congratulations of the Society.

On the Wednesday evening His Majesty gave a dinner party in Windsor Castle to about a hundred guests, representing the agricultural interests of the country and including several members of the Council of the Society. The King, in a short and informal speech of welcome to his guests, expressed the hope that " this historic occasion would be the beginning of a new era, when Agriculture would come into its own."

On Thursday the weather was less pleasant, there being many heavy showers ; unfortunately it was at its worst for the visit of Queen Mary, who arrived in the afternoon. An interesting event was the presentation of the Society's Long Service Medals, by the Earl of Athlone, to a group of farm workers. The first recipient was Edwin Woolgar, who had completed 64 years of service on the Sussex estate of Sir Merrik Burrell.

On Friday Their Majesties returned, this time accompanied by the Princesses Elizabeth and Margaret Rose, and the Royal Party had a most enthusiastic reception from the great crowd. Among the guests at luncheon were the Lord Mayor of London and the Lady Mayoress, the Duke of Norfolk (President of the National Federation of Young Farmers'

Clubs) and Earl De La Warr, President of the Board of Education. His Majesty presented his Cup to the owner of the champion animal in the Young Farmers' Cattle Show. He also presented the Gold Challenge Cup to the Scottish team which had won the International Cattle Judging contest. After the parade of Young Farmers' live stock the Royal Party visited various sections of the Show. At the Old Farm a veteran farrier, who was introduced to Their Majesties, was able to show them a certificate, signed by Queen Victoria, which he had been awarded at the Jubilee Show.

On Saturday there was again broken weather, but the crowd, which exceeded 30,000, was the largest of the week ; it included many organised parties of school children, some 2,000 in all. The Show was visited by His Royal Highness the Duke of Kent.

Among the long list of events in the daily programme in the main ring, the outstanding attraction to the farmer was the Cattle Parade, which was the greatest spectacle of its kind, and the most convincing proof of the skill of British breeders, that had ever been seen. From the historical point of view the parade of ancient vehicles was of outstanding interest, while, as entertainment, most people would have put first the display of horsemanship by a detachment of the Life Guards.

The total attendance for the five days, excluding members and governors and the large number of official guests, reached the very satisfactory total of 118,036.

Thus with a Show which brought together a wonderful collection of live stock and machines, presented a great deal of material of high eductional interest, and was at the same time a brilliant spectacle, the Society rounded off its first hundred years.

INDEX

His Majesty King George VI, President of the Society 1939, 184 ; at the Centenary Show, 198-9.
Aberdeen Angus Cattle, 39, 43, 190.
Adeane, Charles, C. B., 74, 75-6, 180.
Ages of Animals, Dentition in relation to, 45, 107.
Agricultural Education, 136-42.
—— Exhibit, at the Centenary Show, 194.
—— Research Exhibits at Shows, 180.
Agricultural Science, 118-35.
Agriculture, Elements of, Fream, 140.
Agriculture, English, in 1838, 1-13.
Agriculture of England and Wales, Society's Memoir on, 1878, 166.
Althorp, Lord (3rd Earl Spencer), 14-15, 37.
Amos, C. E., Consulting Engineer, 31, 84-5.
Animal Disease, 104-17.
Animal Diseases Acts, 113-16.
Anthrax, 116.
Anti-Corn-Law League, 13, 15.
Arch, Joseph, 145, 148.
Athlone, Earl of, Foreword. Deputy President 1939, 184.

Battersea, Show at, 1862, 46.
Bedfordshire Farming, 29, 147.
Bell, Patrick, 88.
Biffen, Sir Rowland (Consulting Botanist), 131.
Binder, Corn, 94-6.
Blight, Potato, 128-9.
Board of Agriculture (Old), 1.
—— —— —— Library of, 36.
Bones, as Manure, 4, 146.
Botanical Research, 128-31.

Breed Societies, 57, 61.
British Friesian Cattle, 78, 190.
Bulls, licensing of, 81-2.
Burke, U. Roland (Sir Roland), Honorary Director, 75, 185-6.
Burrell, Sir Merrik, Bt., 82, 183.
Butter-making Competitions, 177.

Caird, Sir James, 166, 167.
Cambridge, Shows at, 26, 63.
—— University confers honorary degrees on Society's officials, 63.
Carlisle Show of 1880, 54.
Carruthers, William (Consulting Botanist), 129-31.
Cathcart (Earl), 129, 166-7, 169.
Cattle, Breeds of, 8-9, 27-8, 38-9, 51, 60, 77-8, 190.
—— Dairy, Improvement of, 49, 79.
—— Plague, 48, 51, 108, 109-12.
Centenary, The Society's, 184.
—— Show, 184-99.
—— —— Historical Exhibits at, 193-4.
—— —— Implements and Machinery at, 192-3.
—— —— Live Stock at, 186-92.
Charter, The Society's, 25 ; Supplementary, 71.
Cheese-making, 175-6.
—— Competition, 177.
Cider and Perry, 178-9.
Clarke, J. Algernon, 162.
Clarke, Sir Ernest (Secretary), 59, 74, 167, 168.
' Clean Milk ' Competitions, 177-8.
Cobden, Anti-Corn-Law League, 15.
Coleman, Professor John, 162.
Cooper, Sir Richard, 75, 156.

INDEX

Theiler, Sir Arnold, 182.

Thompson, H. S. (Sir Harry S. Meysey), 31, 85, 119–20, 163–4.

Thorold, Sir John, 169.

Threshing Machine, 6, 87–8.

Tile-making Machines, 30, 32, 86–7.

Tithe Commutation, 12.

Tractor, Agricultural, 99–103.

Tuley, Joseph, 40.

Turner, T. B. (Secretary), 185.

Union of Agricultural Labourers, 145, 148.

Unsoundness in Horses, 45, 49, 61, 109.

Veterinary College, Royal, 104, 107, 108, 116–17.

Veterinary Inspection of Horses, 45, 49, 61, 109.

Veterinary Science and Animal Disease, 104–17.

Voelcker, Augustus (Consulting Chemist), 120–2, 176.

Voelcker, J. A. (Consulting Chemist), 122, 125.

Wages of Agricultural Workers, 11, 145, 148, 149, 158.

War Work, The Society's, 180–2.

Warwickshire, Farming in, 148, 155.

Way, J. T. (Consulting Chemist), 119–20.

Webb, Jonas, 26, 28–9, 40, 42, 47.

Wheat, 6, 23–4, 32–3, 150, 152, 155.

'Whisky Money' for Agricultural Education, 139–40.

Wilson, Sir Jacob (Honorary Director), 50, 55, 61–2, 66–9.

Wiltshire, Farming in, 157.

Windsor, Shows at, 39, 59–61, 185–99.

Woburn, Society's Experiments at, 119, 122–8, 130, 132, 134.

Woodlands Competition, 173–4.

Working Dairy at Show, 176–7.

Yorkshire, Farming in, 147, 151–2, 155, 158.

Young Farmers' Clubs, 82, 190, 191.